The Mathematics of Chip-firing

Discrete Mathematics and Its Applications

Series Editors
Miklos Bona
Donald L. Kreher
Patrice Ossona de Mendez
Douglas West

Representation Theory of Symmetric Groups
Pierre-Loïc Méliot

Advanced Number Theory with Applications
Richard A. Mollin

Handbook of Linear Algebra, Second Edition
Leslie Hogben

Combinatorics, Second Edition
Nicholas A. Loehr

Handbook of Discrete and Computational Geometry, Third Edition
C. Toth, Jacob E. Goodman and Joseph O'Rourke

Handbook of Discrete and Combinatorial Mathematics, Second Edition
Kenneth H. Rosen

Crossing Numbers of Graphs
Marcus Schaefer

Graph Searching Games and Probabilistic Methods
Anthony Bonato and Paweł Prałat

Handbook of Geometric Constraint Systems Principles
Meera Sitharam, Audrey St. John, and Jessica Sidman,

Additive Combinatorics
Béla Bajnok

Algorithmics of Nonuniformity:
Tools and Paradigms
Micha Hofri and Hosam Mahmoud

Extremal Finite Set Theory
Daniel Gerbner and Balazs Patkos

https://www.crcpress.com/Discrete-Mathematics-and-Its-Applications/book-series/CHDISMTHA
PP?page=1&order=dtitle&size=12&view=list&status=published,forthcoming

The Mathematics of Chip-firing

Caroline J. Klivans

CRC Press
Taylor & Francis Group
Boca Raton London New York

CRC Press is an imprint of the
Taylor & Francis Group, an **informa** business

CRC Press
Taylor & Francis Group
6000 Broken Sound Parkway NW, Suite 300
Boca Raton, FL 33487-2742

© 2019 by Taylor & Francis Group, LLC
CRC Press is an imprint of Taylor & Francis Group, an Informa business

No claim to original U.S. Government works

Printed on acid-free paper
Version Date: 20181018

International Standard Book Number-13: 978-1-138-63409-1 (Hardback)

Library of Congress Cataloging-in-Publication Data

Names: Klivans, Caroline J., 1977- author.
Title: The mathematics of chip-firing / Caroline J. Klivans.
Description: Boca Raton : CRC Press, Taylor & Francis Group, 2018. | Includes bibliographical references.
Identifiers: LCCN 2018034117 | ISBN 9781138634091
Subjects: LCSH: Graph theory. | Combinatorial analysis. | Abelian groups. | Sequences (Mathematics)
Classification: LCC QA166 .K544 2018 | DDC 511/.5--dc23
LC record available at https://lccn.loc.gov/2018034117

Visit the Taylor & Francis Web site at
http://www.taylorandfrancis.com

and the CRC Press Web site at
http://www.crcpress.com

To Audrey, Aaron, and Pedro.

Contents

Preface

Chip-firing processes are discrete dynamical systems. A commodity (chips, sand, dollars) is exchanged between sites of a network according to simple local rules. Although governed by local rules, the long-term global behavior of the system reveals unexpected properties.

Physicists introduced these systems under the name the abelian sandpile model, in the context of self-organized criticality. Sandpiles are modeled by a height function along a discrete two-dimensional grid. Similarly, combinatorialists envision stacks of chips sitting at vertices of a graph. From either perspective the basic dynamics are the same. At any time step, if the height of the sand (or number of chips) exceeds a certain threshold, then the sand disperses to neighboring sites. The dispersion of sand at one site may in turn cause another site to reach threshold and itself disperse. Repeated dispersion or avalanching is at the heart of chip-firing dynamics. The addition of a single grain of sand can cause large scale avalanching. One may expect that, over time, the model would simply even out to a level amount of sand at all sites. Instead, the final configurations that result from avalanching turn out to be highly intricate.

Part I of the book covers the fundamentals of chip-firing. Chapter 1 begins with a brief introduction to chip-firing including an extended example of the basic dynamic. Chapter 2 presents the details of chip-firing dynamics such as the abelian property, stabilization and criticality. A strong connection to combinatorics is made in Chapter 3 with the result that the number of distinct long-term stable configurations is equal to the number of spanning trees of the graph. Furthermore, we present Merino's Theorem which refines this enumeration and has spurred much interest in the combinatorics of chip-firing due to its connection to face numbers.

Chapters 4 and 5 continue with early perspectives on chip-firing. Chapter 4 treats the sandpile group, a finite abelian group naturally associated to chip configurations. Chapter 5 discusses pattern formation, including the identity element of the sandpile group. It is here that one finds the captivating fractal behavior of chip-firing models.

Part II of the text presents more general frameworks for chip-firing. Chapter 6 builds from the observation that chip-firing is a form of discrete diffusion governed by the graph Laplacian. Appropriately generalizing the graph Laplacian to other operators yields new systems but with similarly nice properties. In this setting, chip-firing can be seen as an energy minimizing system.

Chapter 7 introduces chip-firing in higher dimensions. Instead of chips on vertices of a graph, the higher-dimensional model consists of flows on cells of a topological complex. Chip-firing in higher dimensions brings in the theory of cellular spanning trees and combinatorial Laplacians.

Chapter 8 introduces a direction motivated by algebraic geometry. Interpreting a graph as a combinatorial analogue of an algebraic curve, chip configurations can be thought of as divisors on curves. The sandpile group plays the role of the Picard or Jacobian group. A highlight of this area is the Riemann–Roch Theorem for graphs. Chapter 8 also includes chip-firing from the perspective of arithmetic geometry and the connection to a two-variable zeta-function.

Finally, Chapter 9 considers chip-firing from the perspective of combinatorial commutative algebra. Chip-firing moves are encoded in a binomial ideal known as the toppling ideal. Also presented here is a monomial initial ideal, the tree ideal of a graph. We will see that the standard monomials of the tree ideal are in bijection with the long-term stable configurations of the chip-firing system.

Acknowledgments

Thank you to the many people who read pages, provided insights and feedback: Ben Braun, Carina Curto, Anton Dochtermann, Art Duval, Pedro Felzenszwalb, Luis Garcia Puente, Sam Hopkins, Lionel Levine, Kevin Marshall, Jeremy Martin, Alex McDonough, Kathryn Nyman, Mackenzie Unger, and Emily Winn. An additional thank you to Carina without whom this book would have never been started and to Pedro for constant help along the way.

Part I

Fundamentals

Chapter 1

Introduction

1.1   A brief introduction

We start with a brief introduction to the basic dynamics of the chip-firing process in order to give a sense of the fundamental action. The overarching idea to keep in mind is: simple local rules leading to complex global behavior.

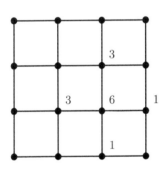

Consider the 4×4 grid to the left. The grid can be thought of as an abstract graphical network or as the discretization of a planar surface. Four sites are populated with an initial non-zero value – one might visualize a stack of chips, grains of sand or an amount of currency.

Two sites connected by an edge are called neighbors. In the chip-firing process, the action of a site will depend on only two things, the value at the site itself and the number of neighbors it has. If the value at a site is at least as large as the number of neighbors it has, it will give one unit of value to each neighbor. This happens regardless of the neighbors' values.

In our example, the site with value 6 has four neighbors. Because its value is greater than its number of neighbors, the site will disperse some of its value. Dispersion is egalitarian and one unit of value is passed to each neighbor. When a site passes value to its neighbors the site *fires*.

Below on the left, the site with value 6 fires once. The result is shown on the right. The value of the initial site has decreased by 4 and the value at each neighbor has increased by 1.

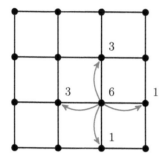

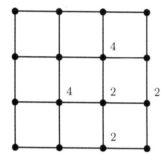

The firing of one site may in turn cause other sites to fire. In our example, two sites with four neighbors now have value 4 and will therefore share with their neighbors. Note that in firing, they will each return one unit of value back to the first site that fired.

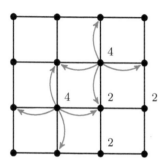

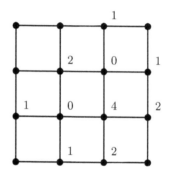

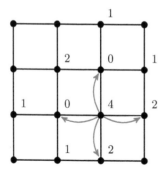

The result of firing these two sites is the same regardless of which site fires first or if they fire simultaneously. The result is shown above and to the right.

In the resulting configuration, the original site that fired has now regained enough value that it will fire again. Depending on the initial configuration, two sites might pass value back and forth in this manner many times.

Although some value has returned to the original location, we see that overall the configuration is more spread out than in the initial configuration. For example, sites along the top and left boundaries now have non-zero value.

Boundary sites have fewer neighbors and hence a different threshold for firing. In our running example, below and to the left, two boundary sites currently have value 3. These sites have only three neighbors and hence they will fire.

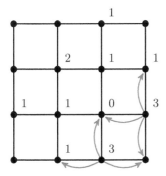 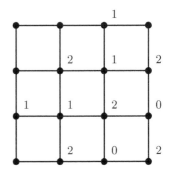

The result of firing both of these locations is that the corner site will have value 2. A value of 2 is enough to allow the corner site to fire.

After firing the corner, the system settles into a configuration where no sites fire. A configuration in which no site can fire is called *stable*.

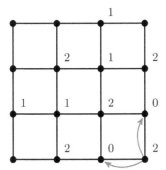 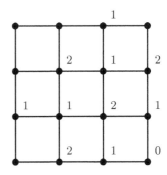

Many immediate questions arise in considering this *chip-firing* process. For example, when two sites can both fire, does it matter in which order they fire? In our example it did not. We will see that the answer is always no. Chip-firing processes satisfy a certain *confluence property* and the final configuration is invariant to the order of firings.

Does the chip-firing process always terminate? This answer depends on the initial number of chips and their distribution. In our example, the process stopped, but envision the grid above with an initial configuration consisting of 4 chips at each site. No matter how the chips are distributed, some site will always have at least 4 chips and therefore the process does not stop.

What if we have different boundary rules? For example, what if chips are able to leave the system? Instead of chips, picture sand on a table top that falls to the floor as it reaches the edge. With this or any other kind of *sink*, we will see that the chip-firing process always terminates. And, in this case, due to the confluence property, every initial configuration has a unique terminating configuration.

We will investigate how many different terminating configurations there are for a given graph – it turns out to be the same as the number of spanning trees of the graph. We will alter the firing rules to allow for asymmetric firing and different thresholds for firing. And we will fire across networks more general than graphs including cell complexes in higher dimensions.

Chip-firing induces a natural algebraic structure, a finite abelian group called the sandpile group. The sandpile group plays a prominent role in a perspective of chip-firing motivated by algebraic geometry. By regarding a graph as an analog of a Riemann surface, we will interpret chip configurations as divisors on curves and firing as linear equivalence. In this setting, the sandpile group will play the role of the Picard group or Jacobian of the surface.

Before all of this, let us look at another, much larger, example that demonstrates some of the beautiful and highly structured mathematics that arises from such simple dynamics.

Instead of the 4 × 4 grid, imagine a very large grid – a much finer discretization of a planar surface. Suppose that initially the grid is empty and imagine adding chips to the center site.

After adding four chips, the center site will fire. After adding sixteen chips and firing four times from the center location, the neighbors of the center will be able to fire. After twelve firings from the center, the next ring of sites will be able to fire. With each firing, some chips will be sent further from the center, while some will be sent back towards the center. If the grid is large enough compared to the total number of chips added, the process will eventually terminate. It is natural to anticipate that the chips will simply disperse across the surface, leveling out as evenly as possible. Instead, this setup leads to a fascinating final configuration.

Figure 1.1 shows the final configuration for an initial configuration of 20 million chips at the center of a grid. In the final configuration, each location in the grid can have between 0 and 3 chips. If any site had more chips, it would fire. The displayed visualization color-codes the number of chips at each site. Thus the triangular red regions represent a collection of sites all of which came to rest with precisely three chips.

These fractal-like images have helped spur research into chip-firing dynamics. Interestingly, although we can observe very clear patterns, we are only beginning to prove formal statements about such configurations.

1.2 Origins and History

Chip-firing processes have been introduced into the literature a number of times from various communities. We give a brief overview of the different origins.

1.2.1 The abelian sandpile model

Bak, Tang and Wiesenfeld introduced the notion of *self-organized criticality* as an explanation of the underlying cause of the widespread occurrence of $1/f$ noise [BTW88]. The authors considered the chip-firing process on a grid, thought of both as a cellular automaton and a discrete dynamical system.

From an initial configuration, the process is run until a stable configuration is reached, such as in our small example or more interestingly as in Figure 1.1. From a dynamical systems point of view, these *critical configurations* are attractors. An initial configuration eventually settles at a unique critical configuration even when the system is initiated far from the equilibrium.

The critical configurations are a subset of the stable configurations. But, their stability is considered not robust in that the addition of just a few chips, sometimes even just one new chip, can cause a large amount of successive firing. The restabilization of a configuration after the addition of chips is known as *avalanching*. The distribution of lengths of avalanches resulting from adding a single chip satisfies a power law with exponent -1, hence the $1/f$ noise.

Bak, Tang and Wiesenfeld actually considered more general setups than just chip-firing, such as systems of pendulums. After their original work, Bak in particular championed the concept of self-organized criticality as a fundamental principle explaining a wide array of phenomena including forest fires, stock markets and neural activity in the brain. For

The values $0 - 3$ have been coded as:

Color	Number of chips
⬛	0
⬜	1
⬜	2
⬛	3

FIGURE 1.1: The stabilization of 20 million chips at the origin.

more in this direction, the interested reader is pointed to Bak's book *How Nature Works: the science of self-organized criticality* [Bak96].

Dhar [Dha90], expanding on the work of Bak, Tang and Wiesenfeld, studied the self-organized criticality phenomenon on more general networks with more general firing rules. His *abelian sandpile model* is formulated as a Markov chain. The states of the chain are the stable configurations of a fixed network. Transitions occur by adding a new chip to the network and restabilizing. Dhar characterized the critical configurations as the recurrent configurations of this chain. He furthermore showed that the limiting distribution of the system is uniform.

Dhar's contributions to the theory of chip-firing are extensive and appear throughout the text. His notes *Theoretical studies of self-organized criticality* [Dha06] give an excellent introduction to abelian models and self-organized criticality and discuss many of their properties which are not covered in this book.

1.2.2 A combinatorial game

Another origin of chip-firing processes is as a combinatorial game. Perhaps the first study in this direction was of a balancing game by Spencer [Spe86]. An initial configuration of N chips is placed at the origin of the one-dimensional grid. At each time step, the chips are distributed as evenly as possible to neighboring sites. Spencer's motivation in studying the system was to balance a collection of vectors in the max norm.

Anderson, Lovász, Shor, Spencer, Tardos and Winograd [ALS+89] studied the dynamics of firing N chips from the origin of the one-dimensional grid from a more combinatorial perspective. They observed the confluent property of the chip-firing rule and studied the final stable configurations reached based on the parity of N. They were also able to determine the precise number of fires required at each site until stabilization in terms of simple binomial coefficients.

Following this work, Björner, Lovász and Shor [BLS91] broadened the domain of consideration to arbitrary undirected graphs and arbitrary initial configurations. Among other important contributions, they characterized the regimes of finite versus infinite processes based on the total number of chips of the initial configuration. Their work also brought in the theory of greedoids and antimatroids. Björner and Lovász [BL92] further considered chip-firing dynamics on directed graphs.

Biggs [Big99a] studied chip-firing processes on graphs as a certain dollar game. Here an initial configuration consists of both positive and negative values. Both firing and reverse-firing moves are used to attempt to get all sites to a positive value. The famous Pentagon problem can be seen as a variant of the dollar game. And similar to the Pentagon problem, which requires a positive total value to be winnable, the dollar game requires a total value greater than the genus of the graph to be winnable.

Biggs also explicitly considered a natural group structure that can be given to the collection of critical configurations of a graph, which he referred to as the critical group of a graph. Bacher, de la Harpe and Nagnibeda [BdlHN97] introduced this group in connection to the lattices of cuts and flows of the graph, while Lorenzini [Lor91] called the group the group of components. In the physics literature we see the name the sandpile group.

1.2.3   Abstract rewriting systems

One of the most fundamental properties of the chip-firing process is the notion of *confluence*. For chip-firing this manifests itself as follows: the order in which firings occur does not influence the final state of the system.

The idea of confluence is not new and has been well explored in the computer science literature on state space models, especially with respect to *Petri nets, vector addition systems* and *abstract rewriting systems*. In the context of abstract rewriting systems, various confluence properties, e.g. weak, local, semi, and global confluence, are considered.

The chip-firing process satisfies a local confluence sometimes known as the Church–Rosser property [CR36]: if, at some time step, two different sites can both fire then the two sites can be fired in succession, in either order, and the resulting configuration is the same. The uniqueness of the final configuration, from a fixed initial configuration, is then a consequence of Newman's Lemma [New42]. Newman's Lemma guarantees that a terminating process which satisfies the local confluency property has a unique terminating state. This property of having a unique terminating state is referred to as global confluence.

Some of the most recent explorations in chip-firing have been extensions of the chip-firing process which do not satisfy the local property yet do satisfy global confluence.

Chapter 2

Chip-firing on Finite Graphs

While chip-firing had much of its origins on infinite graphs, it has since been extended to many other domains. In this chapter, we will focus on finite undirected graphs as the core model for the fundamental properties of chip-firing.

2.1 The chip-firing process

Let $G = (V, E)$ be a finite connected undirected simple graph on $|V| = n$ vertices.

Definition 2.1.1.

- A *chip configuration* for G is any non-negative integer vector

$$\mathbf{c} = (c_1, c_2, \ldots, c_n) \in \mathbb{Z}_{\geq 0}^n.$$

  We interpret $\mathbf{c}$ as recording the number of chips (grains of sand, amount of commodity) located at each vertex of the graph.

- A vertex v is said to be *ready to fire* if the number of chips at v is at least the number of neighbors of v:

$$v \text{ is ready to fire if } c_v \geq \deg(v).$$

- A vertex *fires* by sending one chip to each of its neighbors. This results in a new configuration $\mathbf{c}'$ where c_v' has decreased by $\deg(v)$ and $\deg(v)$ entries have all increased by 1. A *legal fire* is one in which no entry of $\mathbf{c}'$ is negative, namely the vertex that fired was ready to fire.

- A configuration $\mathbf{c}$ is *stable* if no vertex is ready to fire:

$$c_v < \deg(v) \text{ for all } v \in V.$$

- The *chip-firing process* on a finite graph G starts with an initial chip configuration **c**. At each step, a vertex that is ready to fire is selected and fired. Firing a vertex may cause other vertices to become ready to fire. If, at any stage, a stable configuration is reached, the process stops.

Although the chip-firing process is formally defined over a graph with vertices, we will sometimes use the language networks and sites.

Example 2.1.2. Let G be the 3×3 grid graph, as shown in Figure 2.1. Reading top left to bottom right, the initial chip configuration is $(1, 2, 0, 1, 4, 1, 0, 0, 0)$. In the initial configuration, only the center site is ready to fire. After the center site fires, the top middle site is then ready to fire. The top middle site fires again causing a new site to be ready to fire. Eventually the process stops at the configuration $(1, 1, 1, 1, 2, 2, 1, 0, 0)$. In this final configuration, every site has fewer chips than its degree.

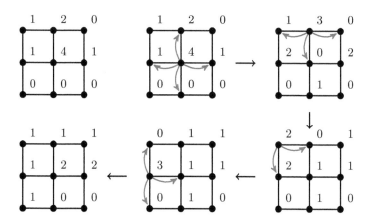

FIGURE 2.1: An unstable initial chip configuration and the subsequent configurations of the chip-firing process.

Example 2.1.3. Again let G be the 3×3 grid graph. Consider the initial configuration $(0, 2, 1, 1, 2, 2, 1, 1, 2)$ shown at the top of Figure 2.2. With this initial configuration, at each stage of the chip-firing process, there is precisely one site ready to fire. After nine firing moves the configuration returns to the initial configuration. In particular, each site fires exactly once before the configuration returns to the initial configuration. Thus this initial configuration never stabilizes and instead results in an infinite chip-firing process.

We remark that in this example, the initial configuration is the out-degree sequence of an acyclic orientation of G. As we will see in the next section (see Theorem 2.3.6) the outdegree sequence of an acyclic orientation always yields a configuration that does not stabilize.

2.1.1 The graph Laplacian

Chip-firing moves can be compactly expressed via the graph Laplacian.

Definition 2.1.4. Let $G = (V, E)$ be a graph on n vertices. The *graph Laplacian* $\Delta(G)$ is the $n \times n$ matrix given by

$$\Delta_{ij} = \begin{cases} -1 & i \neq j \text{ and } \{v_i, v_j\} \in E \\ \deg(v_i) & i = j \\ 0 & \text{otherwise.} \end{cases}$$

Let A be the $n \times n$ adjacency matrix of G and let D be the $n \times n$ diagonal matrix with diagonal given by the degree sequence of G. Then the above definition can be written as

$$\Delta(G) = D - A.$$

The graph Laplacian can alternatively be defined in terms of the oriented incidence matrix of G.

Definition 2.1.5. Let $G = (V, E)$ be a finite oriented graph with $|V| = n$ and $|E| = m$. The *oriented incidence matrix* $\partial(G)$ of G is the $n \times m$ matrix given by

$$\partial(G)_{ve} = \begin{cases} 1 & e = (v, w) \\ -1 & e = (w, v) \\ 0 & \text{otherwise.} \end{cases}$$

Considering G as a one-dimensional simplicial complex, as we will have motivation to do in subsequent chapters, $\partial(G)$ is the standard simplicial boundary map from edges to vertices. Also, when the context is clear, we will suppress the G both from the Laplacian and the incidence matrix.

The graph Laplacian of G is equivalently defined as

$$\Delta = \partial\partial^T,$$

where ∂ is the oriented incidence matrix with respect to any orientation of G.

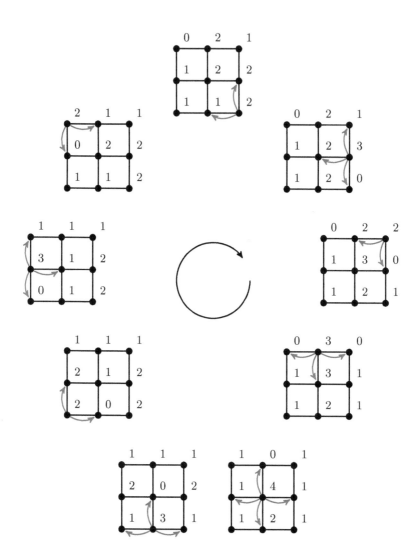

FIGURE 2.2: An unstable initial configuration given by the outdegrees of an acyclic orientation. After all nine sites fire, the process returns to the initial configuration.

Firing a site $v_i \in V(G)$ corresponds to subtracting the ith row of the graph Laplacian from the current configuration. Namely, if a configuration $\mathbf{c}'$ is obtained from the configuration $\mathbf{c}$ by firing at node v_i, then

$$\mathbf{c}' = \mathbf{c} - \Delta e_i,$$

where e_i is the ith standard basis vector in $\mathbb{R}^n$.

In this context, we are representing a chip configuration as a column vector. It is not necessary to use the transpose of the Laplacian here because the Laplacian is symmetric and we choose not to use the transpose for the sake of notational simplicity.

Example 2.1.6. Returning to our examples above, the graph Laplacian for the 3×3 grid graph is a 9×9 matrix. Labeling the nodes from top left to bottom right, the graph Laplacian is

$$\Delta(G) = \begin{array}{c} \\ 1 \\ 2 \\ 3 \\ 4 \\ 5 \\ 6 \\ 7 \\ 8 \\ 9 \end{array} \begin{array}{c} \begin{array}{ccccccccc} 1 & 2 & 3 & 4 & 5 & 6 & 7 & 8 & 9 \end{array} \\ \left(\begin{array}{ccccccccc} 2 & -1 & 0 & -1 & 0 & 0 & 0 & 0 & 0 \\ -1 & 3 & -1 & 0 & -1 & 0 & 0 & 0 & 0 \\ 0 & -1 & 2 & 0 & 0 & -1 & 0 & 0 & 0 \\ -1 & 0 & 0 & 3 & -1 & 0 & -1 & 0 & 0 \\ 0 & -1 & 0 & -1 & 4 & -1 & 0 & -1 & 0 \\ 0 & 0 & -1 & 0 & -1 & 3 & 0 & 0 & -1 \\ -1 & 0 & 0 & -1 & 0 & 0 & 2 & -1 & 0 \\ 0 & 0 & 0 & 0 & -1 & 0 & -1 & 3 & -1 \\ 0 & 0 & 0 & 0 & 0 & -1 & 0 & -1 & 2 \end{array} \right) \end{array}.$$

In Example 2.1.2, the first firing, at site 5, corresponds to

$$(1, 2, 0, 1, 4, 1, 0, 0, 0)^T - \Delta(G)e_5 = (1, 3, 0, 2, 0, 2, 0, 1, 0)^T.$$

In terms of the Laplacian, a configuration $\mathbf{c}$ is stable if

$$\forall i, \ \mathbf{c} - \Delta e_i \not\geq \mathbf{0},$$

where $\mathbf{0}$ is the all zeros configuration.

Thought of as the discretization of the analytic Laplace operator, the graph Laplacian takes on the form:

$$(\Delta f)_{(v)} = \sum_{u \in E} (f(u) - f(v)),$$

for functions $f : V \to \mathbb{R}$. From this perspective, it is quite natural to see chip-firing as a discrete form of diffusion. The interesting behavior of chip-firing comes from an important change to diffusion – the integrality constraints (sites have an integer number of chips).

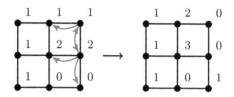

FIGURE 2.3: A legal cluster-fire.

2.1.2   Cluster-fires

Here we introduce a different type of firing move in which multiple sites fire simultaneously. This is also efficiently expressed in terms of the graph Laplacian.

Let $S \subseteq V$ be a non-empty subset of vertices of a graph G. Let χ_S be the characteristic vector of S, i.e. for $S = \{v_{i_1}, v_{i_2}, \ldots, v_{i_k}\}$,

$$\chi_S = \sum_{j=1}^{k} e_{i_j}.$$

Definition 2.1.7. Let G be a graph with Laplacian matrix $\Delta(G)$ and let $S \subseteq V$ be a non-empty subset of vertices of G. The *cluster-fire* of S from a configuration $\mathbf{c}$ results in the configuration $\mathbf{c}'$ given by:

$$\mathbf{c}' = \mathbf{c} - \Delta(G)\chi_S.$$

A *legal* cluster-fire is one in which no entry of $\mathbf{c}'$ is negative.

Example 2.1.8. In Example 2.1.2, corresponding to Figure 2.1, the final stable configuration is $\mathbf{c} = (1, 1, 1, 1, 2, 2, 1, 0, 0)$. Although no individual site is ready to fire, there are legal cluster-fires that can be performed. For example, in Figure 2.3 the top and middle right most sites (vertices 3 and 6) are simultaneously fired to yield the new configuration $\mathbf{c}' = (1, 2, 0, 1, 3, 0, 1, 0, 1)$,

$$\mathbf{c}' = \mathbf{c} - \Delta(G)\chi_{\{3,6\}}.$$

Neither site 3 nor 6 could legally fire. But, if we imagine momentarily that they were to fire, each site would have to send one chip to the other. The cluster-fire cancels out that one-for-one trade and disperses chips to the remaining neighbors. In Figure 2.3 the canceled chip trade is seen by the double-headed arrow.

2.2 Confluence

At any stage in the chip-firing process, there might be multiple sites ready to fire. What are the consequences of the order in which sites are fired? One of the remarkable properties of chip-firing is that there are no consequences; the order of firing does not matter.

If a configuration $\mathbf{c}'$ can be obtained from a configuration $\mathbf{c}$ in a single firing move, we write

$$\mathbf{c} \to \mathbf{c}'$$

Definition 2.2.1. A configuration $\mathbf{d}$ is said to be *reachable* from a configuration $\mathbf{c}$ if there exists a sequence of legal firing moves starting at $\mathbf{c}$ and yielding $\mathbf{d}$:

$$\mathbf{c} = \mathbf{c}_1 \to \mathbf{c}_2 \to \mathbf{c}_3 \to \cdots \to \mathbf{c}_m = \mathbf{d}.$$

The next result gives *the* fundamental property of the chip-firing process. It is based on a simple confluence property.

Theorem 2.2.2 (Commutativity of the Chip-Firing Process).

1. **Local Confluence** *(the diamond property[1]) Suppose $\mathbf{c}_1$ and $\mathbf{c}_2$ are two configurations on a graph G which are both reachable from a configuration $\mathbf{c}$ after one firing. Then there exists a common configuration $\mathbf{d}$ reachable from both $\mathbf{c}_1$ and $\mathbf{c}_2$ after a single firing.*

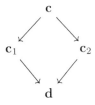

2. **Global Confluence** *(Uniqueness) If a stable configuration $\mathbf{c}_s$ is reachable from a configuration $\mathbf{c}$ after a finite number of legal fires, then $\mathbf{c}_s$ is the unique stable configuration reachable from $\mathbf{c}$.*

Proof. 1. Let configurations $\mathbf{c}_1$ and $\mathbf{c}_2$ be reachable from $\mathbf{c}$ after one firing. Suppose the vertices of G are labeled such that $\mathbf{c}_1$ is the

---

[1]Some sources use the terminology diamond property for a weaker confluence, allowing $\mathbf{d}$ to simply be reachable from $\mathbf{c}_1$ and $\mathbf{c}_2$, not necessarily after a single step; see e.g. [vL90, Chapter 6].

result of firing v_1 and $\mathbf{c}_2$ is the result of firing v_2. Hence sites v_1 and v_2 are both ready to fire in $\mathbf{c}$. Firing site v_1 (resp. v_2) can only increase the number of chips at v_2 (resp. v_1) hence if v_2 was ready to fire before firing site v_1, it is still ready to fire after firing site v_1. Thus $\mathbf{d} = \mathbf{c} - \Delta(G)\chi_{\{v_1,v_2\}}$ is reachable from both $\mathbf{c}_1$ and $\mathbf{c}_2$.

2. In the finite case, repeatedly applying the commutativity of part (1) shows that the final configuration is fixed regardless of the order of firings.

$\square$

By firing sites in different orders, the chip-firing process can take multiple paths to the final stable configuration. Theorem 2.2.2 shows that the final stable configuration is unique regardless of the choices made at each step. The proof shows that more can be said than the statement of the theorem. Not only does the order of legal fires not matter:

- The length of any stabilizing sequence is the same.

- The number of times each site fires in any stabilizing sequence is the same.

Example 2.2.3. Again let G be the 3×3 grid graph. Consider the initial configuration shown in Figure 2.4. Reading top left to bottom right, the initial configuration is $(1, 3, 2, 0, 0, 2, 0, 0, 0)$. In this initial configuration, multiple sites are ready to fire – both the top middle and top right sites. At each step of the chip-firing process, one site that is ready to fire is chosen and fired. Figure 2.4 shows the multiple paths that result from different choices of which site to fire at each step. From each configuration, the consequence of a blue fire is down and to the right, green fires yield the configurations down and to the left. Regardless of the choices made, the final stable configuration is the same.

Consider the sequences of fires represented by the far left and far right paths from the initial to final configuration in Figure 2.4. If we record which sites fired in the order that they fired, we have:

$$2, 1, 3, 6, 3, 2 \text{ and } 3, 6, 2, 3, 1, 2.$$

As multi-sets they are the same and the final configuration is:

$$\mathbf{c}_s = (1, 3, 2, 0, 0, 2, 0, 0, 0)^T - \Delta(G) \cdot (e_1 + 2e_2 + 2e_3 + e_6)^T.$$

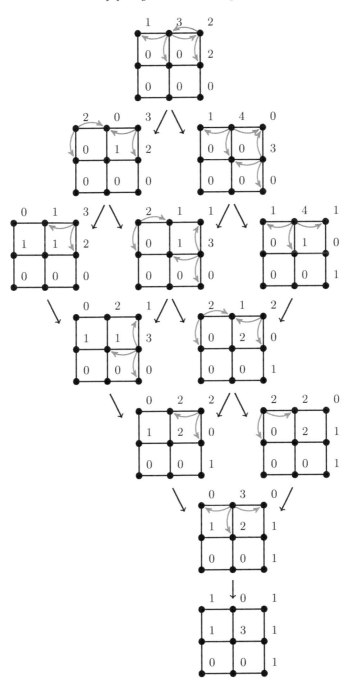

FIGURE 2.4: The multiple paths possible by firing sites in different orders.

The conclusions of Theorem 2.2.2 also imply that with respect to the final state, the chip-firing process could have been equivalently defined with the following firing rule:

At every time step, all sites that can legally fire do so.

2.3 Stabilization

Above we saw that if an initial configuration eventually reaches a stable configuration, then the stable configuration is unique. In their early work on chip-firing games, Björner, Lovász and Shor addressed the question of which initial configurations stabilize. They showed that configurations fall into three regimes depending only on the total number of chips. We follow their presentation here starting with an observation for infinite processes.

Lemma 2.3.1. *If an initial configuration* **c** *on a graph G never reaches a stable configuration, then every vertex of G fires infinitely often in the chip-firing process starting at* **c**.

Proof. If the chip-firing process does not stabilize then there must be some vertex v that fires infinitely often. Every firing of v sends a chip to each neighbor of v. Each neighbor is ready to fire after accumulating some finite number of chips. Therefore each neighbor of v must also fire infinitely often. If the graph is connected, iterating this argument yields that every vertex must fire infinitely often. □

Lemma 2.3.2. *If, from an initial configuration* **c** *on a graph G, every vertex of G can legally fire at least once, then the chip-firing process from* **c** *will never stabilize.*

Proof. Suppose every vertex of G has fired at least once and **c** does eventually stabilize. Consider the vertex v which stopped firing first. Then every other vertex of G fired after the last firing of v. In particular, all neighbors of v fired since the last firing of v. But then v would have gained at least $\deg(v)$ many chips and the configuration would not be stable. □

Corollary 2.3.3. *The chip-firing process starting from an initial configuration* **c** *is finite if and only if there exists some vertex which never fires during the stabilization of* **c**.

Example 2.3.4. In Example 2.1.3, the chip firing process is infinite. Each of the nine vertices fire exactly once before the process returns to the initial configuration.

Example 2.3.5. In Example 2.1.2, which does stabilize, only four sites fire. Sites $1, 3, 6, 7, 8$ and 9 do not fire before stabilization.

We are now ready to analyze how the chip-firing process can evolve based on the number of chips in the initial configuration.

Theorem 2.3.6 ([BLS91]). *Let G be a finite connected graph with n vertices and m edges. Consider configurations* **c** *on G such that $\sum_i c_i = N$, i.e.* **c** *has N chips.*

1. *If $N > 2m - n$ then the chip-firing process with initial configuration* **c** *is infinite.*

2. *If $m \leq N \leq 2m - n$ then there exists an initial configuration which stabilizes and also one which does not.*

3. *If $N < m$ then the chip-firing process with initial configuration* **c** *is finite.*

Proof. Let G be a finite connected graph with n vertices and m edges. Let **c** be a chip configuration with N chips.

1. Suppose $N > 2m - n$. Recall that for any simple graph, $\sum_{v \in V} \deg(v) = 2m$. Then by the pigeon hole principle, there must always exist a vertex with at least $\deg(v)$ chips.

2. Suppose $m \leq N \leq 2m - n$. For $N \leq 2m - n$, the configuration with at most $\deg(v) - 1$ chips at each vertex v is itself stable. For $N = m$, we construct a non-stabilizing initial configuration. First, consider an acyclic orientation $\mathcal{O}(G)$ of G. Let **c** be the initial configuration with $\text{outdeg}_{\mathcal{O}(G)}(v)$ chips at each vertex v. Note that in any acyclic orientation there must exist a source vertex. Hence there is a vertex v in our initial configuration **c** with precisely $\deg(v)$ chips. Fire vertex v and reverse the orientation of all edges incident to v. The resulting orientation is also acyclic and the number of chips at each vertex is again equal to the outdegree. Thus there must exist a source vertex and the process can be repeated indefinitely.

3. Suppose $N < m$. For each edge e in the graph, consider the first chip that fires across e. Associate this first chip to the edge e so that in all subsequent firings the chip either does not move or traverses e. The initial configuration had less chips than edges, thus there is some edge with no associated chip. Therefore there are at least two vertices that never fire. By Corollary 2.3.3, the configuration must stabilize.

□

Example 2.3.7. In the graph of Figures 2.1 and 2.4, there are $n = 9$ vertices and $m = 12$ edges. Theorem 2.3.6 gives that any configuration with $N > 15$ chips will never terminate. For the range $12 \leq N \leq 15$ some initial configurations will terminate while others will not. And finally, any initial configuration with $N < 12$ chips will always stabilize.

In Figures 2.1 and 2.4 the total number of chips are less than 12 and our two initial configurations did indeed stabilize.

The initial configuration of Figure 2.2 has exactly 12 chips and does not stabilize. The example of Figure 2.2 is derived from the construction in part (2) of the proof of Theorem 2.3.6. Figure 2.5 shows the initial acyclic orientation used. The number of chips at each site is equal to the outdegree. Reversing the orientation of all edges incident to the firing vertex gives the acyclic orientation associated to the configuration resulting from the firing.

FIGURE 2.5: The acyclic orientation used to generate the initial configuration of the non-stabilizing example seen in Figure 2.2.

2.4 Toppling time

Suppose we have an initial configuration that does eventually stabilize. As noted just after Theorem 2.2.2, the length of any sequence to stabilization is the same, where the length is measured as the number of individual site fires. What can be said about this length? How long does the chip-firing process take to stabilize? Is it an efficient method? As we have seen, firing has an avalanching effect – one site toppling can cause others and then itself to topple again many times.

Example 2.4.1. Consider a large grid graph and an initial configuration consisting of 78 chips at the origin and no chips elsewhere. Figure 2.6 shows the total number of times each site fires in the stabilization of this initial configuration. For example, the origin will fire a total of 25 times during stabilization. The origin has degree 4, hence we know that the site will need to fire at least 19 times ($= 78/4$). The data shows that the origin will fire an additional 6 times, all of which must be initiated by chips returning to the origin after having fired away. Indeed, 6 chip-firings require a total accumulation of 24 chips, or about a third of all the chips return to the origin.

0	1	2	1	0
1	4	8	4	1
2	8	25	8	2
1	4	8	4	1
0	1	2	1	0

FIGURE 2.6: The number of times each site fires in the stabilization of 78 chips at the origin. All unlabeled sites do not fire.

A more refined notion of toppling time will be investigated in Section 3.8 on avalanche polynomials, where one considers adding a single additional chip to certain special configurations. Here we will be concerned with the total time from initial to final configuration in a finite process.

Tardos showed that this length cannot be too long by giving a polynomial bound (in the number of vertices) on the toppling time. Tardos' bound on the length of any stabilization sequence is given in terms of the number of vertices, the number of edges, and the diameter of the graph.

Definition 2.4.2. For a graph $G = (V, E)$, the *distance* $d(v, w)$ between two nodes v and w, is the length of the shortest path between them. The *diameter* of a graph is the maximum over all pairs of vertices of the distance between them:

$$\text{diam}(G) = \max\{d(v, w) \mid v, w \in V\}.$$

Theorem 2.4.3 ([Tar88]). *Let G be a finite undirected graph with n vertices, m edges, and diameter $d = \text{diam}(G)$. Then any initial configuration on G which eventually stabilizes in the chip-firing process will stabilize within $2mnd$ fires.*

In order to prove the theorem, we first need the following lemma.

Lemma 2.4.4. *For an initial configuration with N chips and two neighboring vertices u and v, at any point in the chip-firing process, the number of times that u has fired cannot differ from the number of times that v has fired by more than N.*

Proof. Suppose that at some point in the chip-firing process, vertex u has fired a times and v has fired b times with $a < b$. Let H_a be the subgraph consisting of all vertices that have fired at most a times. Consider the partitioning of the graph into $(H_a, V \setminus H_a)$.

During the chip-firing process, by construction, H_a has gained chips from $(V \setminus H_a)$ via all edges that bridge the two components. Since the total number of chips is N, H_a could not have gained more than N chips total, so it could not have gained more than N chips across a single bridging edge, hence the difference $a - b$ must be at most N. □

Proof. (Of Theorem 2.4.3) Suppose we have an initial configuration consisting of N chips that eventually stabilizes. By Lemma 2.3.2, there must be some vertex v that never fires in the stabilization process. By Lemma 2.4.4, the maximum number of times a neighbor of v can fire is N and the total number of times any vertex can fire is $N \cdot \text{diam}(G)$. Since the configuration stabilizes, N must be less than or equal to $2m - n \leq 2m$ by Theorem 2.3.6, and the total number of fires is bounded by $2mnd$. □

Example 2.4.5. The diameter of the 3×3 grid graph is 4 which is the length of the shortest path between two corner vertices. Thus Theorem 2.4.3 implies that any initial configuration on the grid graph that does stabilize will do so in no more than $2 \cdot 12 \cdot 9 \cdot 4 = 864$ firings. Note that this refers to an initial configuration with no more than 24 chips!

An alternative bound on the toppling time, given in terms of Laplacian eigenvalues, is given in [BLS91].

In Chapter 6, we will investigate chip-firing on directed graphs. Toppling time is one aspect that changes significantly for directed graphs. In the directed setting, toppling can take exponential time, see Theorem 6.5.15.

2.5   Stabilization with a sink

In Theorem 2.3.6 above, we saw that based on the total number of chips in a configuration, the chip-firing process may stabilize or may continue indefinitely. Here we consider a natural modification of the process which guarantees that *all* initial configurations will eventually terminate. In particular we introduce a sink vertex.

Recall one of our motivating imageries: sand falling onto a table top. Initially, the sand may pile up in the center of the table, eventually avalanching to create various smaller sandpiles along the surface of the table top. As more and more sand is added to the table, eventually sand will fall off the edges of the table. If we simply allow this sand to disappear from the system, then regardless of the amount of sand we drop onto the table, eventually the system will stabilize by losing as much sand as necessary off the edge of the table.

In terms of chip-firing, the table top example is represented by a finite grid. The sink is a single extra vertex connected to every site on the boundary of the grid. The sink, however, is governed by different dynamics than any other site.

Let $G = (V, E)$ be a finite graph on $n + 1$ vertices and let $q \in V$ be a distinguished vertex which we refer to as the *sink*. A chip configuration on a graph with a sink is an integer vector

$$\mathbf{c} = (c_1, c_2, \ldots, c_n, c_q) \text{ such that } c_i \geq 0 \text{ for all } i \neq q.$$

No requirement is placed on the value at c_q.

A configuration $\mathbf{c}$ on a graph G with a sink q is *stable* if

$$\mathbf{c}_v < \deg(v) \text{ for all non-sink vertices } v \neq q.$$

There are two subtly different processes that can be defined in reference to chip-firing with a sink. This depends on whether or not the sink is allowed to fire. We are careful to distinguish them below.

Definition 2.5.1. The *chip-firing process on a graph with a sink*

- (in which the sink does not fire) starts with an initial chip configuration **c** and at each step selects and fires a non-sink vertex that
  is ready to fire. If, at any step, a stable configuration is reached,
  then the process stops.

- (in which the sink does fire) starts with an initial chip configuration
  c and at each step selects and fires a non-sink node that is ready
  to fire. If, at any step, a stable configuration is reached, then (and
  only then) the sink vertex is fired regardless of the value of the
  configuration at the sink.

Therefore, the chip-firing process on a graph with a sink follows the
same dynamics as before except with regard to the sink vertex. In the
first case there will be one vertex of the graph that is simply never
allowed to fire. We will see shortly that in this process, *all initial configurations eventually stabilize.*

In the second case, when no non-sink vertex can fire, the sink is necessarily fired – regardless of whether or not the sink has more chips than
neighbors. When the sink fires, the neighbors of the sink each receive one
additional chip and the value at the sink decreases by the degree of the
sink. Thus the value of a chip configuration at the sink may be negative.
By construction, in this system, even though stable configurations are
reached, *the chip-firing process never terminates.*

The second variant of chip-firing in which the sink fires is often referred to as the *dollar game* as investigated by Biggs [Big99a]. In this
context, the graph represents a network of economies. When one site
has enough funds, it purchases from others sites thus distributing funds.
The sink represents the central bank which only distributes funds to kick-
start a stalled economy, and is allowed, of course, to go into arbitrary
debt.

Proposition 2.5.2. *Let G be a finite graph with a sink q. Every initial
configuration on G eventually reaches a stable configuration under each
of the two processes in Definition 2.5.1.*

Proof. Suppose that there exists an initial configuration that does not
stabilize. As above, this implies that some site v must fire infinitely many
times. All neighbors of v must then also fire infinitely many times. By
considering a path from v to the sink q, we see that some neighbor of q
must fire infinitely many times. However, every firing of a site adjacent
to the sink strictly decreases the total number of chips on non-sink sites.
As the initial configuration had only finitely many chips to start, this
cannot happen infinitely often.   □

The sink guarantees stabilization of all initial configurations, but also introduces an often not desired degree of freedom. For this reason, one typically normalizes chip configurations on graphs with a sink. Given an initial configuration **c** on the non-sink vertices, set the value of the sink vertex to

$$c_q = -\sum_{v \neq q} c_v.$$

With this convention, the total sum of any configuration is always zero. Note that, in general, the total sum of chips is an invariant of the chip-firing process – chips are neither created nor destroyed.

Due to this normalization, chip configurations for a graph with a sink are often displayed with only the values of the non-sink vertices, as in Figure 2.7.

Example 2.5.3. Figure 2.7 shows a stabilization sequence for an initial configuration on $G = C_4$, the four cycle. In the initial configuration, the value at the sink is implicitly assumed to equal -4. At the end of the stabilization process, if the sink is allowed to fire, then the process would return to the initial configuration.

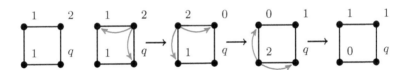

FIGURE 2.7: Stabilization with a sink.

For graphs with a sink, we will often work with a reduced graph Laplacian. Let Δ be the Laplacian of a graph G with sink q. Then the reduced graph Laplacian of G with respect to q is the matrix Δ_q obtained from Δ by deleting the row and column corresponding to q. While Δ is a singular form with the kernel spanned by the all ones vector, Δ_q is non-singular.[2]

Every configuration stabilizes in the chip-firing process on a graph with sink. What effect does the sink have on the toppling time? A number of bounds have been given related to various graphical properties, see e.g. [CE02] and [BS13]. Here we state a bound related to the effective

---

[2]Throughout the text we assume all graphs are connected. In general, the rank of the Laplacian is equal to the number of vertices minus the number of connected components.

resistance of a graph. In comparison to Theorem 2.4.3, the effective resistance R in Proposition 2.5.5 is always bounded above by the diameter of a graph.

Definition 2.5.4. The *effective resistance* $R_{v,w}$ between vertices v and w in a graph G with sink q is given by

$$R_{v,w} = (e_v - e_w)^T \Delta(G)^{-1}(e_v - e_w),$$

where $\Delta(G)^{-1}$ denotes the $n \times n$ matrix formed by setting the first $(n-1) \times (n-1)$ rows and columns equal to the inverse of the reduced graph Laplacian with respect to q and setting each entry of the last row and column equal to zero.

The matrix $\Delta(G)^{-1}$ is a *pseudo-inverse* of $\Delta(G)$ and the effective resistance is derived from Kirchhoff's laws when considering a graph as an electrical network.

The next proposition on toppling times is stated in terms of *chip moves* where firing a vertex v causes $\deg(v)$ chip moves.

Proposition 2.5.5 ([HLM$^+$08]). *Let G be a finite undirected graph with a sink, m edges, and maximum effective resistance R between any vertex and the sink. For any initial chip configuration $\mathbf{c}$ with N chips on the non-sink vertices, the number of chip moves until stabilization is bounded by*

$$2mNR.$$

The proof of this proposition uses the theory of expected hitting times of simple random walks and would take us too far afield of our present context; see [HLM$^+$08] for details.

2.6 Long-term stable configurations

For the remainder of the chapter, if not otherwise stated, we assume all graphs have a sink. Under this assumption, we now know that any initial configuration will eventually stabilize. What is the nature of the stable configurations? It is easy to understand *all* possible stable configurations – every node must simply have fewer chips than its degree. Within these, we will see that there are special families of stable configurations.

The structure and behavior of these special classes has been a major motivation for much work in chip-firing, especially in connection to the notions of self-organized criticality and reduced divisors.

2.6.1 Criticality

Definition 2.6.1. A configuration **c** is *critical* if it is

1. Stable.

2. Reachable from a sufficiently large initial configuration **b**, where a configuration is sufficiently large if every non-sink vertex is ready to fire:
$$b_v > \deg(v) \ \forall v \neq q.$$

Informally, the critical configurations are the configurations that one actually sees when the chip-firing process is run from a generic initial configuration.

Again, there are many stable configurations. Every configuration **c** such that
$$0 \leq c_v < \deg(v) \ \forall \, v \neq q$$
is stable. Disregarding the sink or assuming that configurations have been normalized to a fixed total number of chips, that gives
$$\prod_{v \neq q} \deg(v)$$

many stable configurations. It turns out however that not all stable configurations are reachable from a large initial configuration: not all stable configurations are critical.

Consider, for example, the all zeros configuration. Envision attempting to reach the all zeros configuration from a large initial configuration. Reaching the all zeros configuration is equivalent to having all chips eventually land in the sink. One quickly realizes that for most graphs it is not possible to push all chips to the sink using chip-firing moves, unless you started with a very small and very special initial configuration.

Example 2.6.2. Consider the graph $K_4 \backslash e$, the complete graph on four vertices with one edge removed. Suppose we have chosen one of the vertices with degree two to be the sink, labeled q in the figure below.

The product of the degrees of non-sink vertices is 18. On the other hand, with this choice of sink, $K_4 \backslash e$ has a total of 8 critical configurations. The critical configurations are listed below (leaving off the value of the sink vertex which is assumed to equal the negative of the sum of the values seen).

We will see that the total number of critical configurations does not change with the choice of sink. The critical configurations themselves, however, may change.

Note that $(0,0,0)$ is not a critical configuration of the example.

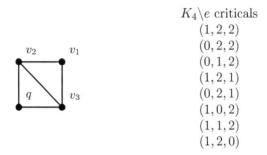

$$K_4 \backslash e \text{ criticals}$$
$$(1,2,2)$$
$$(0,2,2)$$
$$(0,1,2)$$
$$(1,2,1)$$
$$(0,2,1)$$
$$(1,0,2)$$
$$(1,1,2)$$
$$(1,2,0)$$

We have given one definition of critical configurations. It was based on firing from a large initial configuration. There are many known equivalent formulations of critical configurations. In order to state them, we first need two definitions.

The *stabilization of a configuration* $\mathbf{c}$,

$$\text{stab}(\mathbf{c})$$

is the unique stable configuration reachable from $\mathbf{c}$ after a sequence of legal chip-firing moves.

For a graph G, the unique maximal stable configuration $\mathbf{c}_{\max}(G)$ is the degree sequence of G minus 1 in each entry,

$$\mathbf{c}_{\max}(G) = \deg(G) - \mathbf{1}.$$

Theorem 2.6.3. *Let G be a finite connected simple graph with a sink vertex q and let Δ be the graph Laplacian of G. The following are equivalent for configurations $\mathbf{c}$ on G.*

1. $\mathbf{c}$ *is critical:*
   $\mathbf{c} = \text{stab}(\mathbf{b})$ *for some $\mathbf{b}$ with $b_i \geq \Delta_{ii}$ for all $i \neq q$.*

2. $\mathbf{c} = \text{stab}(\mathbf{c} + \mathbf{b})$ *for some $\mathbf{b}$ with $b_i > 0$ for all $i \neq q$.*

3. $\mathbf{c} = \text{stab}(\mathbf{c} + N\mathbf{b})$ *for some $\mathbf{b}$ with $b_i > 0$ for all $i \neq q$ and every integer $N \geq 1$.*

4. $\mathbf{c} = \text{stab}(\mathbf{c} + (-\Delta(q)))$ *where $\Delta(q)$ is the row of Δ corresponding to the sink vertex.*

5. $\mathbf{c} = \text{stab}(\mathbf{c}_{\max} + \mathbf{b})$ *for some $\mathbf{b}$ with $b_i > 0$ for all $i \neq q$.*

6. $\mathbf{c}$ *is stable and re-occurs infinitely often in the chip-firing process in which the sink fires for some initial configuration.*

7. **c** *is stable and there exists a sequence of fires starting from* **c**, *including firing from the sink if necessary, that returns the configuration to* **c**.

8. **c** *is stable and after firing the sink vertex, every non-sink vertex fires exactly once before restabilizing.*

9. **c** *is both stable and allowed, where a configuration* **c** *is* allowed *if for all subsets of vertices* $I \subseteq (V \setminus q)$ *there exists a vertex* $j \in I$ *such that*

$$c_j \geq \sum_{i \in I \setminus j} -(\Delta_q)_{ij}.$$

We leave the proof of Theorem 2.6.3 as an instructive exercise to the reader.

Example 2.6.4. In Example 2.5.3 an initial configuration $\mathbf{c} = (1, 2, 1)$ on C_4 is stabilized to $\mathrm{stab}(\mathbf{c}) = (1, 1, 0)$. The configuration $(1, 1, 0)$ is critical. It is easily seen to satisfy conditions 4, 6, 7 and 8 of Theorem 2.6.3.

2.6.2 Firing equivalence

In order to understand the structure of critical configurations, we first define an equivalence relation on integer vectors motivated by firing reachability.

Definition 2.6.5. Let G be a graph on n vertices with graph Laplacian Δ. Two integer sequences **c** and **d** in $\mathbb{Z}^n$ are *firing-equivalent* on G, denoted

$$\mathbf{c} \sim \mathbf{d}, \text{ if } \mathbf{d} - \mathbf{c} = \Delta \mathbf{z}$$

for some $\mathbf{z} \in \mathbb{Z}^n$. The equivalence class of **c**, denoted $[\mathbf{c}]$ is the *firing class* of **c**.

If **c** and **d** are normalized configurations with the values at the sink vertex suppressed, then **c** is *firing-equivalent* to **d** if

$$\mathbf{d} - \mathbf{c} = \Delta_q \mathbf{z}$$

for some $\mathbf{z} \in \mathbb{Z}^{n-1}$.

Let ∂_0 be the map which adds together all entries of a configuration. Then algebraically, the quotient $\ker \partial_0 / \mathrm{im}(\Delta)$ is isomorphic to the quotient $\mathbb{Z}^{n-1} / \mathrm{im}(\Delta_q) = \mathrm{coker}(\Delta_q)$ and we refer to firing equivalence over Δ_q instead of Δ. We will work with this algebraic perspective much more in Chapter 4.

Note that the equivalence extends to all integer sequences not just non-negative sequences. If one chip configuration can be reached from another through legal fires then they are equivalent. The converse is not necessarily true – two chip configurations can be equivalent but not obtainable from each other via *legal* firing moves.

The next theorem connects firing-equivalence and critical configurations.

Theorem 2.6.6. *For a finite undirected graph G with Laplacian Δ, there exists a unique critical configuration per firing class (equiv. per element of the* $\mathrm{coker}(\Delta_q)$).

Theorem 2.6.6 appears in Dhar's original work on the abelian sandpile model [Dha90]. A variety of proofs exist in the literature, e.g. in [Big99a], [Gab93], [Gab94] and [CR00]. For our proof, we will use the following two lemmas.

Lemma 2.6.7. *Let Δ_q be the reduced graph Laplacian for a graph with n non-sink vertices. Then there exists an integer vector $\mathbf{z} \in \mathbb{Z}^n$ such that $\Delta_q \mathbf{z} > \mathbf{0}$.*

Proof. Let $\mathbf{1}$ be the all ones vector. Let $\mathbf{g}$ be defined by

$$\mathbf{g} = \Delta_q^{-1} \mathbf{1}.$$

The entries of $\mathbf{g}$ are necessarily rational as such let $g_i = \frac{a_i}{b_i}$. Define $\lambda = b_1 b_2 \cdots b_n$ in order to clear denominators. Then $\mathbf{z} = \lambda \mathbf{g}$ will satisfy the statement of the lemma:

$$\Delta_q \mathbf{z} = \Delta_q \lambda \mathbf{g} = \lambda \Delta_q \mathbf{g} = \lambda \Delta_q \Delta_q^{-1} \mathbf{1} = \lambda \mathbf{1} \geq \mathbf{1}.$$

$\square$

As we will see in Chapter 6, reduced graph Laplacians are M-matrices and hence have many nice properties. In particular, we will be able to conclude that $\mathbf{g}$ itself is positive because $\mathbf{1}$ is positive, Δ_q^{-1} is non-negative (by virtue of being an M-matrix), and Δ_q is non-singular. Multiplying by $\mathbf{1}$ has the affect of taking column sums and the column sum of a non-negative non-singular matrix cannot be 0. Therefore the $\mathbf{z}$ in Lemma 2.6.7 is in fact positive.

Lemma 2.6.8. *Suppose $\mathbf{c}$ and $\mathbf{d}$ are two firing equivalent chip configurations; i.e. $\mathbf{d} - \mathbf{c} = \Delta_q \mathbf{z}$ for some $\mathbf{z}$. Then there exists a common configuration $\mathbf{b}$ that legally fires to both $\mathbf{c}$ and $\mathbf{d}$.*

Proof. Break **z** into its positive and negative parts: let $I, J \subseteq [n]$ be defined as $I = \{i \mid z_i \geq 0\}$ and $J = \{j \mid z_j < 0\}$. Then

$$\mathbf{c} - \sum_J z_j \Delta e_j = \mathbf{d} + \sum_I z_i \Delta_q e_i,$$

and

$$\mathbf{b} = \mathbf{c} - \sum_J z_j \Delta_q e_j$$

legally fires to both **c** and **d**. □

Proof. (Of Theorem 2.6.6) First we prove existence. Let [**c**] be a firing equivalence class of G. If **c** is sufficiently large ($c_i \geq \deg_i$) for all i, then we are done as stab(**c**) is critical and an element of [**c**]. If **c** is not sufficiently large, then by Lemma 2.6.7 there is an integer vector **z** such that $\Delta_q \mathbf{z}$ is all positive. For any integer $t > 0$ let $\mathbf{c}_t$ be defined as

$$\mathbf{c}_t = \mathbf{c} + t\Delta_q \mathbf{z}.$$

For large enough t, $\mathbf{c}_t$ must be sufficiently large. Then stab($\mathbf{c}_t$) is critical and an element of [**c**].

Uniqueness follows from Lemma 2.6.8: Suppose **c** and **d** are both critical and firing equivalent. Then there exists a configuration **b** which legally fires to both. But this presents a contradiction since both **c** and **d** are stable by virtue of being critical whereas the stabilization of **b** is unique by Theorem 2.2.2. □

Hence, there is one critical configuration per firing class; equivalently there is one critical configuration per element of the cokernel of the reduced graph Laplacian. In general this means there are far fewer critical configurations than stable configurations. In particular, we will see in the next chapter in Theorem 3.1.3:

The number of critical configurations of a graph G is equal to the number of spanning trees of G.

2.6.3 Superstability

Recall from above that a cluster-fire is the simultaneous firing of multiple sites. How are the dynamics changed if one allows for cluster fires as opposed to only single site firings? At first glance, we might feel there is no difference between the two. Starting from a configuration **c** and sequentially firing a sequence of sites results in the same configuration as firing all the sites simultaneously. The difference arises in considering *legal* fires.

Recall that from an initial configuration **c**, firing a collection of sites S is a *legal* cluster-fire if

$$\mathbf{c} - \Delta\chi_S \geq 0.$$

If we perform a sequence of individual legal fires, then the corresponding cluster-fire will also be legal. It is the converse that fails to hold. *There are legal cluster-fires that are not equivalent to a sequence of individual legal fires.*

Example 2.6.9. Consider the configurations in Figure 2.8.

- The configuration on the left is not stable. The site with 2 chips has degree 2 and thus this site can legally fire.

- The configuration in the middle is stable, every site has fewer chips than its degree. Firing either site individually would leave a value of -1. However, we can cluster-fire the two sites with 2 chips. Firing the two sites simultaneously leaves both with value 0, effectively canceling the chip transfer between the two.

- The configuration on the right has no legal fires and no legal cluster-fires.

With this distinction in mind, define superstable configurations as those in which there are no legal cluster-fires.

Definition 2.6.10. A configuration **c** is *superstable* if

$$\forall S \subseteq V, \quad \mathbf{c} - \Delta\chi_S \not\geq \mathbf{0}.$$

Namely, there are no legal cluster-fires from **c**.

We could also consider configurations **c** such that for all integer vectors **z**, $\mathbf{c} - \Delta\mathbf{z} \not\geq 0$. Namely, we could allow multi cluster-fires. For a finite connected graph, we gain nothing new. A configuration with no legal cluster-fires also has no legal multi cluster-fires. We will see in Chapter 6 that this distinction is important for more general graphs.

As with critical configurations, the superstable configurations can also be interpreted as long-term stable configurations under this extended firing operation. Superstable configurations are also unique per firing class. To see this, we first make a connection to energy minimization.

FIGURE 2.8: The configuration on the left is not stable, the site with 2 chips can legally fire. The middle configuration is stable since no individual site can fire. It is not superstable because the two sites with 2 chips could cluster-fire. The configuration on the right is superstable, there are no legal cluster-fires.

2.6.4 Energy minimization

The chip-firing process starts with an initial configuration and fires it "down" to a stable configuration. Formalizing this idea, we define the *energy* of a configuration. Each chip-firing move lowers the energy of a configuration. If we allow for cluster-fires this results in a minimal energy configuration.

The idea of chip-firing as energy minimization is first due to Baker and Shokrieh [BS13]. We follow [GK15] which develops the theory in a broader context.

Definition 2.6.11. Let G be a graph with sink q and let Δ_q be the reduced graph Laplacian for G with respect to the sink. For a chip configuration $\mathbf{c}$, define the *energy* of $\mathbf{c}$, $E(\mathbf{c})$, as:

$$E(\mathbf{c}) = \|\Delta_q^{-1}\mathbf{c}\|_2^2,$$

where $\|v\|_2^2 = v \cdot v$.

Energy Minimization Problem: Given a chip configuration $\mathbf{c}$, determine a chip configuration of lowest energy that is firing equivalent to $\mathbf{c}$. Namely, find a solution to:

$$\operatorname*{argmin}_{\substack{\mathbf{d} \sim \mathbf{c} \\ \mathbf{d} \geq 0}} E(\mathbf{d}).$$

We call a solution to the energy minimization problem an *energy minimizer*. Since we are working in a discrete space, energy minimizers necessarily exist. We will see that there is a unique energy minimizer per firing equivalence class.

The result will require two lemmas. The first concerns legal versus not legal firing moves.

Given $\mathbf{z} \in \mathbb{Z}^n$ define $\mathbf{z}^+ \in \mathbb{Z}_{\geq 0}^n$ by

$$z_i^+ = \begin{cases} z_i & \text{if } z_i \geq 0 \\ 0 & \text{otherwise.} \end{cases}$$

Similarly, define $\mathbf{z}^- \in \mathbb{Z}_{\leq 0}^n$ by replacing all positive entries of $\mathbf{z}$ with 0.

Lemma 2.6.12. *Let* $\mathbf{c}, \mathbf{d} \geq 0$ *be non-negative chip configurations for a graph G with reduced Laplacian Δ_q. Suppose*

$$\mathbf{d} = \mathbf{c} - \Delta_q \mathbf{z}$$

for some integer vector z. Then the configuration obtained from $\mathbf{c}$ by performing only the firings implied by positive entries of $\mathbf{z}$ is also non-negative:

$$\mathbf{f} = \mathbf{c} - \Delta_q \mathbf{z}^+ \geq 0.$$

Proof. Suppose that $z_i^+ = 0$. Then $-(\Delta_q \mathbf{z}^+)_i \geq 0$. Hence, $\mathbf{f}_i \geq \mathbf{c}_i \geq 0$. On the other hand suppose that $z_i^+ > 0$, then $\mathbf{z}^- = \mathbf{z} - \mathbf{z}^+$ satisfies $z_i^- = 0$ and so $(\Delta_q \mathbf{z}^-)_i \geq 0$, or equivalently $(\Delta_q \mathbf{z})_i \geq (\Delta_q \mathbf{z}^+)_i$ and so $f_i - (\Delta_q \mathbf{z}^+)_i \geq f_i - (\Delta_q \mathbf{z})_i \geq 0$. □

The next lemma expresses the difference in energy of two firing equivalent configurations.

Lemma 2.6.13. *Let $\mathbf{c}$ and $\mathbf{d}$ be chip configurations for a graph G with reduced Laplacian Δ_q. Suppose*

$$\mathbf{d} = \mathbf{c} - \Delta_q \mathbf{z}$$

for some integer vector $\mathbf{z}$, then

$$E(\mathbf{d}) = E(\mathbf{c}) + \mathbf{z}^T \mathbf{z} - 2\mathbf{z}^T \Delta_q^{-1} \mathbf{c}$$
$$= E(\mathbf{c}) - \mathbf{z}^T \mathbf{z} - 2\mathbf{z}^T \Delta_q^{-1} \mathbf{d}.$$

Proof.

$$E(\mathbf{d}) = \|\Delta_q^{-1}(\mathbf{c} - \Delta_q \mathbf{z})\|_2^2$$
$$= \|\Delta_q^{-1} \mathbf{c} - \mathbf{z}\|_2^2$$
$$= \|\Delta_q^{-1} \mathbf{c}\|_2^2 + \mathbf{z}^T \mathbf{z} - 2\mathbf{z}^T \Delta_q^{-1} \mathbf{c}$$
$$= E(\mathbf{c}) + \mathbf{z}^T \mathbf{z} - 2\mathbf{z}^T \Delta_q^{-1} \mathbf{c}$$
$$= E(\mathbf{c}) - \mathbf{z}^T \mathbf{z} - 2\mathbf{z}^T \Delta_q^{-1} \mathbf{d}.$$

□

As mentioned earlier, the reduced graph Laplacian is a special case of an M-matrix. All non-singular M-matrices L have the property that $L^{-1} > 0$ entrywise; see Chapter 6. We will use this fact in our next proof.

Theorem 2.6.14. *Let G be a graph with reduced Laplacian Δ_q. For every chip configuration $\mathbf{c}$ on G, there exists a unique energy minimizer equivalent to $\mathbf{c}$. Namely, for every configuration $\mathbf{c}$, there exists a unique solution to the energy minimization problem.*

Proof. Suppose that $\mathbf{d} \sim \mathbf{c}$ and $\mathbf{b} \sim \mathbf{c}$ with $\mathbf{d}, \mathbf{b} \geq \mathbf{0}$ both energy minimizers. We will show that $\mathbf{d} = \mathbf{b}$. Because $\mathbf{d}$ is equivalent to $\mathbf{b}$, there exists $\mathbf{z}$ such that $\mathbf{d} = \mathbf{b} - \Delta_q \mathbf{z}$ for some $\mathbf{z} \in \mathbb{Z}^n$. By Lemma 2.6.12 we know that $\mathbf{f} = \mathbf{b} - \Delta_q \mathbf{z}^+ \geq 0$. Now, all three are equivalent: $\mathbf{f} \sim \mathbf{b} \sim \mathbf{c}$. By Lemma 2.6.13 we have

$$E(\mathbf{f}) = E(\mathbf{b}) - (\mathbf{z}^+)^T \mathbf{z}^+ - 2(\mathbf{z}^+)^T \Delta_q^{-1} \mathbf{f}.$$

Using that Δ_q^{-1} is a non-negative matrix and $\mathbf{f} \geq 0$, we have that $\Delta_q^{-1} \mathbf{f} \geq 0$. This implies that $-2(\mathbf{z}^+)^T \Delta_q^{-1} \mathbf{f} \leq 0$, and so

$$E(\mathbf{f}) \leq E(\mathbf{b}) - (\mathbf{z}^+)^T \mathbf{z}^+.$$

Since $\mathbf{b}$ is a minimizer it must be that $\mathbf{z}^+ = 0$ or that $\mathbf{z} \leq 0$.

On the other hand, we similarly have

$$E(\mathbf{b}) = E(\mathbf{d}) + \mathbf{z}^T \mathbf{z} - 2\mathbf{z}^T \Delta_q^{-1} \mathbf{b}.$$

Since $\mathbf{z} \leq 0$ this shows that $E(\mathbf{d}) < E(\mathbf{b})$ unless $\mathbf{z} = 0$. $\qquad\square$

We will use the perspective of energy minimization to show uniqueness of superstable configurations by showing that they are in fact the same system of configurations.

Theorem 2.6.15. *For a graph G with reduced Laplacian Δ_q, a chip configuration $\mathbf{c}$ is superstable if and only if it is an energy minimizer.*

Proof. Suppose that $\mathbf{c}$ is superstable and let $\mathbf{d} \sim \mathbf{c}$ with $\mathbf{d} \geq 0$. Then we know that there exists $\mathbf{z} \in \mathbb{Z}^n$ such that

$$\mathbf{d} = \mathbf{c} - \Delta_q \mathbf{z}.$$

By Lemma 2.6.12

$$\mathbf{f} = \mathbf{c} - \Delta_q \mathbf{z}^+ \geq 0,$$

but since $\mathbf{c}$ is superstable then it must be that $\mathbf{z}^+ = 0$.

$$E(\mathbf{d}) = E(\mathbf{c}) + \mathbf{z}^T \mathbf{z} - 2\mathbf{z}^T \Delta_q^{-1} \mathbf{c}$$

and we have that $E(\mathbf{d}) \geq E(\mathbf{c})$.

In the other direction suppose that $\mathbf{c}$ is the energy minimizer. Suppose $\mathbf{c}$ is not superstable. Then this implies there exists $\mathbf{z} \in \mathbb{Z}^n$ with $\mathbf{z} \geq 0$ and $\mathbf{z}$ not identically zero such that

$$\mathbf{d} = \mathbf{c} - \Delta_q \mathbf{z} \geq 0.$$

Since

$$E(\mathbf{d}) = E(\mathbf{c}) - \mathbf{z}^T \mathbf{z} - 2\mathbf{z}^T \Delta_q^{-1} \mathbf{d},$$

this implies that

$$E(\mathbf{d}) \leq E(\mathbf{c}) - \mathbf{z}^T \mathbf{z} < E(\mathbf{c}).$$

However, this contradicts that $\mathbf{c}$ is the energy minimizer and $\mathbf{c}$ must be superstable. $\qquad\square$

Example 2.6.16. Consider $K_4 \setminus e$ as in Examples 2.6.2, 2.6.9 and Figure 2.8.

The three configurations of Figure 2.8 are:

$$
\begin{aligned}
(2,1,1) \qquad &\text{unstable} \\
(0,2,2) \qquad &\text{stable} \\
(0,1,1) \qquad &\text{superstable.}
\end{aligned}
$$

The reduced Laplacian is:

$$
\Delta_q(K_4 \setminus e) = \begin{array}{c} \\ v_1 \\ v_2 \\ v_3 \end{array}
\begin{array}{c}
\begin{array}{ccc} v_1 & v_2 & v_3 \end{array} \\
\left(\begin{array}{ccc}
2 & -1 & -1 \\
-1 & 3 & -1 \\
-1 & -1 & 3
\end{array} \right).
\end{array}
$$

and the corresponding energies $E(\mathbf{c})$ are:

$$
\begin{aligned}
\|\Delta_q^{-1}(2,1,1)\|_2^2 &= 17 \\
\|\Delta_q^{-1}(0,2,2)\|_2^2 &= 12 \\
\|\Delta_q^{-1}(0,1,1)\|_2^2 &= 3.
\end{aligned}
$$

2.6.5 Duality

There is a remarkable and remarkably simple combinatorial duality between the critical and superstable configurations of a graph. Recall the configuration

$$\mathbf{c}_{\max}(G) = \deg(G) - \mathbf{1},$$

the degree sequence of G minus 1 in each entry. The configuration $\mathbf{c}_{\max}(G)$ is the unique maximum stable chip configuration.

Duality here will manifest as complementing via $\mathbf{c}_{\max}$:

$$\text{critical} \longleftrightarrow \text{superstable}$$
$$\mathbf{c} \qquad\qquad \mathbf{c}_{\max} - \mathbf{c}$$

In order to prove the duality between superstable and critical configurations, we need two lemmas.

Lemma 2.6.17. *Let* $\mathbf{c}$ *be an unstable chip configuration and* $\mathbf{d}$ *a stable configuration such that*

$$\mathbf{d} = \mathbf{c} - \sum_{j=1}^{k} \Delta e_{i_j}.$$

Then for every site v *that is unstable in* $\mathbf{c}$, *there exists* j *such that* $i_j = v$.

Proof. Suppose that v is unstable in $\mathbf{c}$:

$$c_v \geq \Delta_{vv}.$$

Further suppose that $i_j \neq v$ for all j. Since the off diagonal entries of Δ are non-positive, we would have

$$\sum_{j=1}^{k} (\Delta e_{i_j})_v \leq 0,$$

and so

$$\mathbf{d}_v \geq \mathbf{c}_v \geq \Delta_{vv},$$

contradicting the fact that $\mathbf{d}$ is stable. □

The next result represents a very important feature of chip-firing. Here we will use it only as a technical lemma, but we will return to it in Chapter 5. The result is known as the *least action principle*. It states that, in essence, the chip-firing process does the least amount of work possible in moving from an unstable to stable configuration via the Laplacian.

Lemma 2.6.18 (Least Action Principle). *Suppose* $\mathbf{d} = \text{stab}(\mathbf{c})$, *i.e.* $\mathbf{d}$ *is stable and results from a sequence of legal fires from* $\mathbf{c}$. *Suppose* $\mathbf{z}$ *is such that*

$$\mathbf{d} = \mathbf{c} - \Delta \mathbf{z}.$$

If

$$\mathbf{f} = \mathbf{c} - \Delta \mathbf{w}$$

is also stable, then $\mathbf{w} \geq \mathbf{z}$.

Proof. Expand $\mathbf{d}$ as

$$\mathbf{d} = \mathbf{c} - \sum_{j=1}^{k} \Delta e_{i_j}$$

with the e_i indexed so that each intermediate difference as j runs from 1 to k is a legal firing.

Now suppose that $\mathbf{w}$ is not greater than $\mathbf{z}$. Then, this implies there exists $1 \le \ell \le k$ such that $\mathbf{w} = \sum_{j=1}^{\ell-1} e_{i_j} + \tilde{\mathbf{w}}$ with $\tilde{\mathbf{w}} \ge 0$ and $\tilde{w}_{i_\ell} = 0$. However, by our hypothesis $(\mathbf{c} - \sum_{j=1}^{\ell-1})_{i_\ell} \ge \Delta_{i_\ell i_\ell}$. Moreover, $\mathbf{c} - \Delta \mathbf{w} = \mathbf{c} - \sum_{j=1}^{\ell-1} -\Delta\tilde{\mathbf{w}}$. By Lemma 2.6.17 it must be $\tilde{w}_{i_\ell} > 0$ which is a contradiction. $\qquad\square$

Theorem 2.6.19. *For a graph G, a configuration $\mathbf{c}$ on G is superstable if and only if $\mathbf{c}_{\max}(G) - \mathbf{c}$ is critical.*

Proof. Let $\mathbf{c}$ be superstable. The configuration $\mathbf{c}_{\max} -\mathbf{c}$ is necessarily stable since $\mathbf{c}_{\max}$ is the unique componentwise maximal stable configuration.

By Lemma 2.6.7 there exists a vector $\mathbf{z} \ge 0$ such that $(\mathbf{c}_{\max} -\mathbf{c} + \Delta \mathbf{z})_i \ge \Delta_{ii}$ for all i. Set $\mathbf{d} = \mathbf{c}_{\max} -\mathbf{c} + \Delta \mathbf{z}$ so that $d_i \ge \Delta_{ii}$ for all i. Note that since $\mathbf{z} \ge 0$ we can write $\mathbf{z} = \sum_{j=1}^{k} e_{i_j}$. We know that

$$\mathbf{c}_{\max} -\mathbf{c} = \mathbf{d} - \sum_{j=1}^{k} \Delta e_{i_j}.$$

The proof will be complete if we can show there exists a permutation σ of $[k]$ so that

$$\mathbf{d}^\ell = \mathbf{d} - \sum_{j=1}^{\ell} \Delta e_{i_{\sigma(j)}},$$

is such that

$$\mathbf{d}^\ell_{i_{\sigma(\ell+1)}} \ge \Delta_{i_{\sigma(\ell+1)} i_{\sigma(\ell+1)}}, \tag{2.1}$$

for $1 \le \ell \le k - 1$. In words, we need to show that there is a *legal* firing from the configuration $\mathbf{d}$ to the configuration $\mathbf{c}_{\max} -\mathbf{c}$.

We proceed to define the permutation σ inductively. Suppose that we have chosen $\sigma(1), \sigma(2), \ldots, \sigma(r-1)$ with $r \le k$ so that (2.1) holds for $1 \le \ell \le r-2$. We know that

$$\mathbf{d}^{r-1} = \mathbf{c}_{\max} -\mathbf{c} + \Delta\tilde{\mathbf{z}}$$

or equivalently

$$\mathbf{c} - \Delta\tilde{\mathbf{z}} = \mathbf{c}_{\max} -\mathbf{d}^{r-1},$$

where

$$\tilde{\mathbf{z}} = \sum_{j=1}^{k} \Delta e_{i_j} - \sum_{j=1}^{r-1} \Delta e_{i_{\sigma(j)}} \geq 0$$

and $\tilde{\mathbf{z}} \neq 0$. Since $\mathbf{c}$ is superstable we know that there exists a q such that $(\mathbf{c}_{\max} - g^{r-1})_q < 0$ or equivalently $\mathbf{d}_q^{r-1} \geq \Delta_{qq}$. Also, since $\mathbf{c} = \mathbf{d}^{r-1} - (\sum_{j=1}^{k} \Delta e_{i_j} - \sum_{j=1}^{r-1} \Delta e_{i_{\sigma(j)}})$, by Lemma 2.6.17 there exists $1 \leq \sigma(r) \leq k$ such that $\sigma(r) \neq \sigma(j)$ for $j = 1, 2, \ldots, r-1$ such that $i_{\sigma(r)} = q$.

For the other direction, note that we have shown that critical configurations are unique up to firing equivalence class, Theorem 2.6.6. We have also shown that superstable configurations are unique up to firing equivalence class, Theorems 2.6.14 and 2.6.15, and that their duals are critical configurations, Theorem 2.6.19, this is enough to conclude the converse. □

Example 2.6.20. Returning to the running example $K_4 \backslash e$, we list all critical and all superstable configurations below. In this example, $\mathbf{c}_{\max} = (1, 2, 2)$.

	$K_4 \backslash e$ criticals	$K_4 \backslash e$ superstables
	$(1, 2, 2)$	$(0, 0, 0)$
	$(0, 2, 2)$	$(1, 0, 0)$
	$(0, 1, 2)$	$(1, 1, 0)$
	$(1, 2, 1)$	$(0, 0, 1)$
	$(0, 2, 1)$	$(1, 0, 1)$
	$(1, 0, 2)$	$(0, 2, 0)$
	$(1, 1, 2)$	$(0, 1, 0)$
	$(1, 2, 0)$	$(0, 0, 2)$

Recall the discussion that $\mathbf{0}$ is essentially never a critical configuration. On the other hand, $\mathbf{0}$ is always a superstable configuration. Duality then gives that $\mathbf{c}_{\max}$ is always critical.

2.6.6 Structure

The collection of superstable (resp. critical) configurations are highly structured as seen in the next two propositions.

Proposition 2.6.21. *Suppose* $\mathbf{c}$ *is a superstable configuration and* $0 \leq \mathbf{b} \leq \mathbf{c}$, *i.e.* $\mathbf{b}$ *is coordinatewise less than or equal to* $\mathbf{c}$. *Then the configuration* $\mathbf{b}$ *is also superstable. Namely, the set of superstable configurations is componentwise downward closed.*

Proof. Suppose $\mathbf{c}$ is superstable and $\mathbf{b} \leq \mathbf{c}$ is not superstable. Then by definition there must exist an integer sequence $\mathbf{z}$ such that $\mathbf{b} - \Delta\mathbf{z} \geq \mathbf{0}$. But since $(\Delta\mathbf{z})_i \leq b_i \leq c_i$ for all i, it must be that $\mathbf{c} - \Delta\mathbf{z} \geq \mathbf{0}$ contradicting the superstability of $\mathbf{c}$. □

Note that by duality, this shows that critical configurations are componentwise upwards closed to $\mathbf{c}_{\max}$: If $\mathbf{c}$ is critical and $\mathbf{b}$ is such that $\mathbf{c} \leq \mathbf{b} \leq \mathbf{c}_{\max}$ coordinate-wise, then $\mathbf{b}$ must also be critical.

There is a useful homogeneity to the maximal superstable (resp. minimal critical) configurations.

Proposition 2.6.22 ([Big99b], [Mer01]). *Componentwise minimal critical configurations all have the same total number of chips on non-sink vertices.*

Proof. Let $G = (V, E)$ be a graph with sink q. We will argue that the total number of chips (not on q) in a minimal critical configuration is equal to $|E| - \deg(q)$.

Consider starting at the configuration $\mathbf{c}$ and firing the sink vertex q. Label each chip fired from the sink by q.

Since $\mathbf{c}$ is critical, by part 8 of Theorem 2.6.3, there is a firing sequence of the vertices of G which, starting from $\mathbf{c}$ returns to $\mathbf{c}$. Proceed firing according to this sequence. When an unlabeled chip is fired from vertex v, label it v. In firing a site, first fire all labeled chips back to their original positions before firing unlabeled chips. Again because $\mathbf{c}$ is critical and this firing sequence returns back to $\mathbf{c}$, every labeled chip will return to its initial position.

The collection of labeled chips forms a minimal critical configuration; it satisfies the conditions of part 8 of Theorem 2.6.3 by construction and the configuration is componentwise less than $\mathbf{c}$. The total number of labeled chips that are not labeled q is equal to $|E| - \deg(q)$. □

Example 2.6.23. Above we listed all critical and superstable configurations for $K_4 \backslash e$. The minimal criticals and maximal superstables are shown below. The minimal criticals all have $5 - 2 = 3$ chips. The maximal superstables dually have 2 chips.

$K_4 \backslash e$ min criticals	$K_4 \backslash e$ max superstables
$(0, 1, 2)$	$(1, 1, 0)$
$(0, 2, 1)$	$(1, 0, 1)$
$(1, 0, 2)$	$(0, 2, 0)$
$(1, 2, 0)$	$(0, 0, 2)$

Duality allows one to immediately compute the superstable configurations (criticals) given the criticals (superstables). A word of caution is in order though. Let $\mathbf{c}$ be a critical configuration and $\mathbf{s} = \mathbf{c}_{max} - \mathbf{c}$ its dual, then $\mathbf{c}$ and $\mathbf{s}$ are not typically firing equivalent. Thus they do not represent the same element of $\mathrm{coker}_{\mathbb{Z}}(\Delta_q)$. For example, in the previous example, $(1, 0, 1)$ and $(0, 2, 1)$ are dual but they are not firing equivalent.

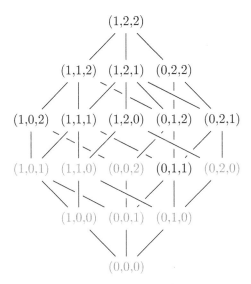

FIGURE 2.9: The poset of positive integer sequences componentwise less than $(1, 2, 2)$. The superstable configurations of $K_4 \setminus e$ are highlighted.

Geometrically, the stable configurations are the integer points inside a box sitting at the origin. Superstable configurations form a downward closed collection above the origin.

Critical configurations form an upward closed collection below the opposite corner. The corner opposite the origin is the configuration $\mathbf{c}_{max}$. Chip-firing duality swaps these two collections by subtracting through the far corner. Figure 2.9 shows the stable configurations of $K_4 \setminus e$ in the componentwise partial order.

One of the original motivations for studying chip-firing processes was the concept of self-organized criticality. Through only local moves, an initial configuration settles to a critical configuration. Critical configurations are stable but only just so. Namely, critical configurations are stable configurations but a small change, say the addition of a few new chips, will cause the system to fire, potentially many times. Consider the maximal configuration c_{max}, if a single chip is added to any site, this will in turn cause every site to fire before stabilizing. Such large chain reactions are referred to as *avalanching*. Intuitively, superstable configurations are the opposite, they are the configurations that are strongly stable. One can add chips to superstable configurations and potentially cause no firings – the resulting configuration remains stable.

2.6.7   Burning

In this section we investigate the question of detecting critical and superstable configurations.

*Given a configuration **c** can we efficiently determine whether or not it is critical? superstable?*

Determining if a configuration **c** is stable is straightforward, we simply check if $c_i < \Delta_{ii}$ for all i. By definition, superstable configurations are those for which there are no legal cluster-fires. Cluster firing simultaneously fires any subset of the vertices. Therefore, a priori, given an initial configuration, one would need to check all possible subsets of the vertices to see if they admit a legal cluster-fire. For a graph on n vertices, this requires 2^n firing checks. Dhar devised a *burning algorithm* which gives a procedure for determining whether or not a configuration is superstable in only n firing checks.

Dhar's burning algorithm has a lively description: Let G be a finite graph with sink q. Let **c** be a chip configuration on $G \setminus q$. Envision the chips as firefighters protecting their location. Imagine a fire burning through the graph – if a vertex v is on fire then the fire spreads along all incident edges towards the neighbors of v. Each firefighter can turn and stop the fire along exactly one edge. Thus, a site is protected as long as there are at least as many firefighters (chips) as there are incident burning edges. If a site has more burning incident edges than firefighters, then the site catches on fire and the fire spreads through towards all of its neighbors.

Theorem 2.6.24. *Let G be a finite graph with sink q. Start a fire at q. A configuration is superstable if and only if every vertex is eventually on fire.*

Dhar's burning algorithm uses this observation to input a configuration **c** and output either a subset of vertices which admit a legal fire or returns that **c** is superstable.

Procedure *Burning Algorithm*(G)
1. Set $A_0 = V$
2. Set $v_0 = q$
3. **for** $1 \leq i \leq n - 1$ do
4.   Let $A_i = A_{i-1} \setminus v_{i-1}$
5.   **if** for all $v \in A_i$, $\mathbf{c}_v \geq \text{outdeg}_{A_i}(v)$ **then** Output A_i.
6.   **else** Let $v_i \in A_i$ be any vertex with $\mathbf{c}_{v_i} < \text{outdeg}_{A_i}(v)$.
7. Output Superstable.

The proof of correctness of the algorithm and justification of the claim that it takes n checks is below:

• If the algorithm outputs $A_i \neq \emptyset$ then the degree requirement is met hence A_i can fire and the initial configuration is not superstable.

• If $\mathbf{c}$ is superstable then by definition, the condition of step 5 can never be satisfied.

• Suppose A_i becomes empty then we claim $\mathbf{c}$ is superstable. Let $U \subseteq V \setminus q$ be some subset of non-sink vertices; we will show that U cannot fire. At first, $U \subset A_1 = V \setminus q$. Since $A_i \to \emptyset$ some $u \in U$ must be removed. Let u be the first node that is removed. Because $U \subset A_1$, $\text{outdeg}_{A_1}(u) < \text{outdeg}_U(u)$. Moreover, $c_u < \text{outdeg}_{A_1}(u) < \text{outdeg}_U(u)$ and therefore U cannot fire.

• The algorithm runs through at most n vertices checking at most n inequalities at each step.

Example 2.6.25. Figure 2.10 shows a configuration on $K_4 \setminus e$. A fire is started at the sink; it is stopped along the bottom and middle paths. The fire does burn through the vertex with 0 chips and then eventually burns through all sites. Therefore the chip configuration is superstable.

FIGURE 2.10: Dhar's burning algorithm for Example 2.6.25.

The burning algorithm can be dually described to detect critical configurations. This is related to the description of critical configurations in terms of firing the sink.

A configuration $\mathbf{c}$ is critical if starting at $\mathbf{c}$ then firing the sink and stabilizing yields $\mathbf{c}$ again:

$$\mathbf{c} = \text{stab}(\mathbf{c} + \Delta e_s).$$

The process of firing the sink adds one chip to each neighbor of the sink vertex. This may add up to n chips to the configuration. Theorem 2.4.3 guarantees that the toppling process will not take more than $2mnd$ steps to stabilize.

Definition 2.6.26. For a graph G with sink q and reduced Laplacian Δ_q, an integer vector $\mathbf{b} \geq 0$ is a *burning configuration* for G if $\mathbf{b}$ is in the integer image of Δ_q:

$$\mathbf{b} = \Delta_q \mathbf{z}$$

for some $\mathbf{z} \in \mathbb{Z}^n$.

Let $\mathbf{b}$ be a burning configuration. The vector

$$\mathbf{z} = \Delta_q^{-1} \mathbf{b}$$

is called the *burning script* for $\mathbf{b}$.

The burning script records the number of times each site fires in the stabilization process caused by the addition of a burning configuration.

The criticality of a configuration can be checked using any burning configuration.

Proposition 2.6.27. *Let $\mathbf{b}$ be a burning configuration for a graph G. A configuration $\mathbf{c}$ on G is critical if and only if*

$$\mathrm{stab}(\mathbf{c} + \mathbf{b}) = \mathbf{c}.$$

Proof. If $\mathrm{stab}(\mathbf{c} + \mathbf{b}) = \mathbf{c}$, then $\mathbf{c}$ is critical by Theorem 2.6.3 part 2. For the other direction, assume $\mathbf{c}$ is critical. By definition, $\mathrm{stab}(\mathbf{c} + \mathbf{b})$ must be stable. If $\mathbf{c}$ is critical, and performing a sequence of topplings on $\mathbf{c}$ results in a stable configuration $\mathbf{x}$, then $\mathbf{x}$ must also be critical by Theorem 2.6.3 part 6. Hence $\mathrm{stab}(\mathbf{c} + \mathbf{b})$ is critical. Furthermore, it is firing equivalent to $\mathbf{c}$ since $\mathbf{b}$ is in the image of the Laplacian. But, critical configurations are unique per equivalence class so $\mathbf{c}$ must equal $\mathrm{stab}(\mathbf{c} + \mathbf{b})$. $\square$

Example 2.6.28. Consider $K_4 \backslash e$:

The reduced graph Laplacian equals

$$\Delta_q = \begin{array}{c} \\ v_1 \\ v_2 \\ v_3 \end{array} \begin{array}{ccc} v_1 & v_2 & v_3 \\ \left(\begin{array}{ccc} 2 & -1 & -1 \\ -1 & 3 & -1 \\ -1 & -1 & 3 \end{array}\right). \end{array}$$

The configuration $\mathbf{b} = (0, 1, 1)$ is a burning configuration. The configuration $\mathbf{b}$ is equal to the product $\Delta_q \mathbf{1}$.

We can check, for example, that the configuration $(1, 2, 2)$ is critical by computing that the stabilization

$$\mathrm{stab}((1, 2, 2) + (0, 1, 1))$$

is equal to $(1, 2, 2)$.

2.7 The sandpile Markov chain

In this last section of the introductory chapter we consider an alternate stochastic viewpoint of the chip-firing process. In this narrative, chip configurations are called *sandpiles* and one envisions grains of sand as opposed to chips at sites of a network. Chip-firing dynamics can then be seen as modeling critical behavior in natural phenomena; see [Dha06].

Once again suppose we have a finite graph G with a sink, hence every initial sandpile (configuration of grains of sand) eventually stabilizes to a unique stable configuration.

2.7.1 Avalanche operators

Definition 2.7.1. The ith *avalanche operator*, A_i is a map on the stable configurations of a graph given by

$$A_i(\mathbf{s}) = \mathrm{stab}(\mathbf{s} + e_i).$$

In terms of the firing dynamics, consider a stable sandpile $\mathbf{s}$. The addition of e_i represents adding a new single grain of sand at site i. This differs from the chip-firing dynamics above where the process starts from a fixed initial configuration and the amount of chips in the system is fixed throughout. Here, we are allowing sand to be added to the graph.

The addition of a grain of sand to site i may cause site i to become unstable and fire. This in turn may cause other sites to become unstable and also fire. $A_i(\mathbf{s})$ equals the resulting stable configuration after all subsequent fires.

In order to prove an abelian property, we first make an observation about stabilization:

$$\mathrm{stab}(\mathbf{c} + \mathbf{z}) = \mathrm{stab}(\mathrm{stab}(\mathbf{c}) + \mathbf{z}) \text{ for any } \mathbf{z} \in \mathbb{N}^n. \tag{2.1}$$

In words, Equation 2.1 says that the same final configuration is reached whether two configurations are first added together and then stabilized, or if one configuration is stabilized, the second is added and then the result is stabilized.

Proposition 2.7.2. *The avalanche operators commute:* $A_i A_j = A_j A_i$.

Proof. This follows from expanding the avalanche operators in terms of the stabilization operator as above. Starting from any stable configuration $\mathbf{s}$, we have:

$$A_i A_j(\mathbf{s}) = \text{stab}(\text{stab}(\mathbf{s} + e_j) + e_i) = \text{stab}(\text{stab}(\mathbf{s}) + e_j + e_i)$$

$$= \text{stab}(\text{stab}(\mathbf{s}) + e_i + e_j) = \text{stab}(\text{stab}(\mathbf{s} + e_i) + e_j) = A_j A_i(\mathbf{s}).$$

$\square$

This abelian property prompted the name for the abelian sandpile model.

Definition 2.7.3. Let $G = (V, E)$ be a finite undirected graph with a sink and let σ be a probability distribution over V. The *abelian sandpile model* is a Markov chain with state space equal to the set of stable configurations of G. Transitions are given by choosing a site v according to σ and performing the avalanche operator A_v.

Example 2.7.4. The four cycle C_4 has a total of 8 stable configurations, all integer vectors componentwise between $(0, 0, 0)$ and $(1, 1, 1)$. Figure 2.11 shows the stable configurations positioned as vertices of the cube. The edge from stable configuration $\mathbf{s}_1$ to $\mathbf{s}_2$ is labeled with the e_i that satisfies $\text{stab}(\mathbf{s}_1 + e_i) = \mathbf{s}_2$.

States $(0, 0, 0)$, $(1, 0, 0)$, $(0, 1, 0)$ and $(0, 0, 1)$ are not recurrent. From the figure, one can see that none of these states are in a directed cycle. The four stable states $(1, 0, 1)$, $(1, 1, 0)$, $(0, 1, 1)$ and $(1, 1, 1)$ are recurrent. These four stable states are precisely the critical configurations of the chip-firing process on C_4.

As a Markov chain, it is natural to ask which stable configurations are recurrent.

Theorem 2.7.5. *The recurrent configurations of the abelian sandpile model are precisely the critical configurations of the chip-firing process.*

Proof. This follows directly from Theorem 2.6.3 part 6. $\square$

In many places in the literature on chip-firing, critical configurations are in fact called *recurrent configurations*.

The sandpile Markov chain stationary distribution is the uniform distribution over recurrent (= critical) configurations. This follows from the presentation in Chapter 4 where recurrent configurations are given the structure of a finite abelian group.

Jerison, Levine and Pike have shown that the mixing time of the sandpile Markov chain has an inverse relationship to the mixing time of a random walk on the underlying graph. The proof uses the theory of random walks on groups; see [JLP15] for details. See also [LW95] for a survey of mixing times of random walks on graphs and the abelian sandpile model.

There is a large body of literature on sandpiles from the probability and statistical physics viewpoint which we do not discuss in this book. See [Jár14] and [Jár18] for an introduction to this area.

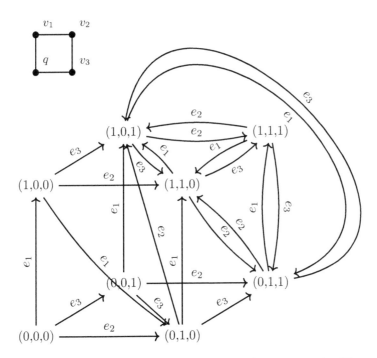

FIGURE 2.11: The sandpile Markov chain for the 4-cycle. There are a total of 8 stable configurations on C_4. Of these, 4 configurations are recurrent states of the Markov chain.

2.8   Exercises

For Exercises 2.8.1 - 2.8.5, let G be the graph shown below:

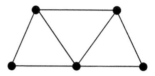

Exercise 2.8.1. *Chip-fire from various initial configurations on G. For each initial configuration* **c**, *find the stabilization or determine that the configuration does not stabilize. In the second case, determine the limiting behavior, i.e. what collection of sites reoccur infinitely often in the chip-firing process starting from* **c**.

Exercise 2.8.2. *Fix an acyclic orientation of G. Let* **c** *be the induced chip configuration given by one less than the outdegree sequence of the orientation. Fire from* **c** *until returning to* **c**.

Exercise 2.8.3. *Fix a sink vertex of G, compute all of the critical and superstable configurations of G with respect to the chosen sink.*

Exercise 2.8.4. *Confirm the superstability of one of the configurations from Exercise 2.8.3 using the burning algorithm.*

Exercise 2.8.5. *Find a burning configuration for G. Confirm the superstability of one of the configurations from Exercise 2.8.3 using the burning configuration.*

Exercise 2.8.6. *Find all superstable configurations of the cycle graph C_n.*

Exercise 2.8.7. *With respect to Theorem 2.3.6, show that in the non-stabilizing regime, with $N > 2m - n$, there can be two different length paths from a configuration back to itself.*

Exercise 2.8.8. *With respect to Tardos' bound on the toppling time, show that this bound is best possible by giving an example of a graph and initial configuration that takes $\Theta(n^4)$ chip-firing moves to stabilize.*

Exercise 2.8.9. *Prove the equivalence of the two given definitions of the graph Laplacian.*

Exercise 2.8.10. *For a graph G with n vertices and c connected components, prove that the rank of $\Delta(G)$ is $n - c$.*

Exercise 2.8.11. *For a connected graph G with sink q, prove that $\Delta(G)_q^{-1}$ has all non-negative entries.*

Exercise 2.8.12. *Suppose that $\mathbf{c}$ is an initial configuration with N chips on a graph with n vertices. Show that if $\mathbf{c}$ eventually stabilizes, then the number of fires until stabilization is at most*

$$\frac{2nN}{\lambda_1},$$

where λ_1 is the smallest non-zero eigenvalue of the graph Laplacian (the algebraic connectivity of the graph).

Exercise 2.8.13. *Prove the equivalence of the characterizations of critical configurations in Theorem 2.6.3.*

Exercise 2.8.14. *Example 2.6.2 gives the critical configurations for $K_4 \backslash e$. Confirm that they are indeed critical configurations by showing that they satisfy one/some of the conditions of Theorem 2.6.3.*

Exercise 2.8.15. *For a graph G with degree sequence d, define the configuration e as follows:*

$$e = d + (d - \mathrm{stab}(d + d)).$$

Prove that a configuration $\mathbf{c}$ is critical if and only if

$$\mathrm{stab}(\mathbf{c} + e) = \mathbf{c}.$$

Exercise 2.8.16. *Prove that for a graph G with sink q, if $\mathbf{c}$ is a critical configuration of G with respect to q, then the extended configuration $\bar{\mathbf{c}}$ with value $\deg(q)$ at vertex q is a nonstabilizing configuration for G.*

Exercise 2.8.17. *The Pentagon Problem.*

Let G be a 5-cycle (a pentagon). Let $\mathbf{c}$ be an initial configuration on G allowing for both positive and negative values but such that the sum of all values over all vertices is positive. If at any time there is a vertex v with a negative value $\mathbf{c}_v$ then the value at v is added to both neighbors of v and the value at v is negated.

Prove that this process terminates, showing in particular, that all vertices can be given a non-negative value through this process.

Exercise 2.8.18. *Prove that the set of configurations reachable from an initial configuration is a lattice.*

Exercise 2.8.19. *Give an example of a graph G and a configuration $\mathbf{c}$ on G such that $\mathbf{c}$ is both critical and superstable.*

Exercise 2.8.20. *Write code to simulate the abelian sandpile model.*

Determine the recurrent configurations for $K_4 \backslash e$.

Confirm the recurrent configurations are precisely the critical configurations.

Chapter 3

Spanning Trees

In this chapter we explore the strong connections between chip-firing and combinatorics. Much of the work in this direction is motivated by the fact that for finite undirected graphs with a sink:

The number of critical configurations of a graph is equal to the number of spanning trees of the graph.

After establishing the equality above, we present two finer enumerative results. First, we look at Merino's Theorem. Merino's Theorem uses a matroid perspective to prove a common recursion between critical configurations and spanning trees. The result shows that the number of critical configurations of a fixed *level* is equal to the number of spanning trees of a fixed *activity*. This leads to an important application of chip-firing: the proof of Stanley's O-conjecture for cographic matroids.

The second result we look at in detail is the Cori–Le Borgne bijection between critical configurations and spanning trees. Many bijections have been formulated between configurations and trees. The Cori–Le Borgne bijection is a refinement of the Merino result in that it is activity/level preserving. The main technique used for this result is a refined burning algorithm.

From the launching point of trees, many other classical combinatorial objects are brought into the chip-firing narrative. As we will see, acyclic orientations align with minimal critical configurations, parking functions are in bijection with superstable configurations, and domino tilings of grids are equinumerous with certain symmetric configurations.

The connection to acyclic orientations will be of particular importance in Chapter 8 on divisor theory where they play an important role in the development of the Riemann–Roch Theorem for graphs.

3.1   Spanning trees

In Chapter 2 we saw that for a finite graph with sink, there are in general far fewer critical configurations than stable configurations.

Theorem 2.6.6 showed that there is a unique critical configuration per equivalence class of the firing equivalence relation. Recall that two configurations are firing equivalent if their difference is in the image of the Laplacian. These equivalence classes are precisely the elements of the integer cokernel of the reduced graph Laplacian, $\mathrm{coker}_{\mathbb{Z}}(\Delta_q)$. The reduced graph Laplacian appeared already in Chapter 2, especially in the proof of Theorem 2.6.6. We repeat the definition here.

Definition 3.1.1. For a graph G on n vertices with Laplacian Δ and any vertex v, the *reduced Laplacian of G with respect to v* is the $(n-1) \times (n-1)$ matrix Δ_v formed from Δ by deleting the row and column corresponding to v.

In our context, when forming a reduced Laplacian for a graph with a sink, if not stated otherwise, we will always delete the row and column corresponding to the sink vertex q.

The full graph Laplacian, Δ, is a singular form – for a connected graph on n vertices, the rank of Δ is $n - 1$. By construction, the kernel of Δ is spanned by the all ones vector $\mathbf{1}$. The reduced graph Laplacian Δ_v is non-singular for any connected graph.

Furthermore, the cardinality of $\mathrm{coker}_{\mathbb{Z}}(\Delta)$ is equal to the index of the lattice generated by the columns of Δ which is the determinant of the reduced Laplacian. By the geometry of numbers, see e.g. [Cas97], each equivalence class is uniquely represented in the fundamental parallelepiped generated by the columns of Δ. Therefore the number of critical configurations is given by:

$$|\text{criticals}| = \det(\Delta_q).$$

The Matrix-Tree Theorem connects this same determinant to the enumeration of spanning trees, see e.g. [Moo70] or [Sta99].

Theorem 3.1.2 (Matrix-Tree Theorem).   *The number of spanning trees, $\tau(G)$, of a connected graph G is given by the determinant of the reduced Laplacian.*

$$\tau(G) = \det(\Delta_v(G)),$$

for any vertex v.

Equivalently, the number of spanning trees of a connected graph is given by the product

$$\tau(G) = \frac{\lambda_1 \lambda_2 \cdots \lambda_{n-1}}{n}$$

where $\lambda_1, \lambda_2, \ldots, \lambda_{n-1}$ *are the non-zero eigenvalues of* Δ.

As a corollary we have:

Theorem 3.1.3. *The number of critical configurations (equiv. the number of superstable configurations) of a connected graph G is equal to the number of spanning trees of G.*

One consequence of the Matrix-Tree Theorem is that for a graph Laplacian Δ, the determinants of the reduced Laplacians Δ_v are the same for all vertices v. Example 3.1.4 shows that the critical (and hence superstable) configurations can change with varying choices of the sink. Although the critical configurations themselves may be different, the *number* of critical configurations is invariant to the choice of sink.

Example 3.1.4. Let $G = K_4 \backslash e$ be the complete graph on 4 vertices minus an edge. Figure 3.1 shows all eight spanning trees of G. Figure 3.2 shows the two possible choices of sink vertex up to isomorphism.

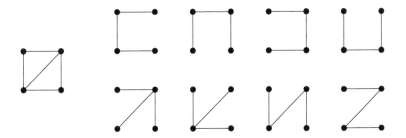

FIGURE 3.1: The eight spanning trees of $K_4 \setminus e$.

Consider the choice of sink as in G_1. The determinant of the reduced Laplacian $\Delta_q(G_1)$, which consists of the first three rows and columns of $\Delta(G_1)$, is equal to 8.

$$\Delta(G_1) = \begin{array}{c} \\ v_1 \\ v_2 \\ v_3 \\ q \end{array} \begin{array}{c} \begin{array}{cccc} v_1 & v_2 & v_3 & q \end{array} \\ \left(\begin{array}{cccc} 2 & -1 & -1 & 0 \\ -1 & 3 & -1 & -1 \\ -1 & -1 & 3 & -1 \\ 0 & -1 & -1 & 2 \end{array} \right) \end{array}.$$

The eigenvalues of $\Delta(G_1)$ are $4, 4, 2, 0$. The spanning tree count can be reaffirmed in the calculation:

$$\frac{4 \cdot 4 \cdot 2}{4} = 8.$$

For any choice of sink vertex, we should find precisely eight critical and eight superstable configurations. Below, the critical and superstable configurations for the two choices of sink are listed. Note that by changing the sink vertex, the entries of the 3-tuples for G_1 do not correspond to the same vertices in G_2. In particular, the critical configurations change by more than just a permutation of entries. For example, $(1, 2, 2)$ is a critical configuration of G_1 but no permutation of $(1, 2, 2)$ is a critical configuration for G_2.

The table below lists each critical configuration with the unique superstable configuration in the same firing equivalence class.

G_1		G_2	
criticals	supers	criticals	supers
$(1,2,2)$	$(1,0,0)$	$(1,1,1)$	$(0,0,0)$
$(0,2,2)$	$(0,0,0)$	$(1,2,1)$	$(0,1,0)$
$(0,1,2)$	$(0,0,1)$	$(0,2,1)$	$(1,0,0)$
$(1,2,1)$	$(1,1,0)$	$(1,2,0)$	$(0,0,1)$
$(0,2,1)$	$(0,1,0)$	$(0,2,0)$	$(0,2,0)$
$(1,0,2)$	$(0,2,0)$	$(0,1,1)$	$(0,1,1)$
$(1,1,2)$	$(1,0,1)$	$(1,1,0)$	$(1,1,0)$
$(1,2,0)$	$(0,0,2)$	$(1,0,1)$	$(1,0,1)$

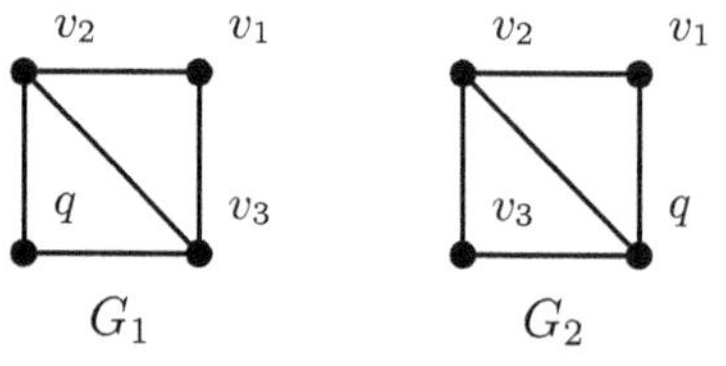

FIGURE 3.2: Varying the choice of sink changes the set of critical configurations.

For each graph, from Theorem 2.6.19, the set of superstable configurations could have alternatively been obtained by dualizing the critical configurations through $\mathbf{c}_{\max}$. For G_1, $\mathbf{c}_{\max} = (1, 2, 2)$ and each superstable configuration can be found as $\mathbf{c}_{\max} - \mathbf{c}$ for some critical configuration $\mathbf{c}$. Such a pairing can be seen in Example 2.6.23.

The conclusion of Theorem 3.1.3 naturally leads one to consider explicit mappings between critical configurations and spanning trees. Many bijections have been constructed between the critical or superstable configurations and spanning trees. The earliest is perhaps by Dhar and Majumdar [MD92] who used a selective burning algorithm. Other associations include those by Biggs and Winkler [BW97], Merino [ML97], Cori and Le Borgne [CLB03], and Chebikin and Pylyavskyy [CP05]. We will look at two in particular.

First, we consider Merino's Theorem which shows an even finer enumeration of critical configurations and spanning trees. Specifically, the result shows that the generating function enumerating critical configurations by the statistic known as the *level* is equal to the generating function enumerating spanning trees by *activity*. The result is achieved by taking on a matroid perspective and establishing a common recursion of both structures. Merino's Theorem also has an important application in the theory of face numbers, presented in Section 3.3.1.

Second, we will see the Cori–Le Borgne bijection. This construction gives a bijective extension of Merino's result. Extending the work of Dhar and Majumdar [MD92] their activity/level preserving mapping is established via a refined burning algorithm.

3.2 Statistics on trees

In the previous section, we saw that critical configurations (and hence also superstable configurations) of a graph G are equinumerous with the spanning trees of G. We will consider two explicit bijections in terms of certain statistics on trees and configurations.

3.2.1 Level

The statistic on chip configurations that we will be concerned with is essentially the total number of chips in the configuration (excluding the sink vertex). The total number of chips is often known as the *weight* of a configuration:

$$\text{wt}(\mathbf{c}) = \sum_{v \neq q} c_v.$$

We require an adjustment of the weight statistic because of the sensitivity of chip configurations to the choice of sink vertex. In considering any connection between critical configurations and spanning trees, this sensitivity to the choice of sink, as demonstrated in Example 3.1.4, is an obvious difficulty. For a graph G, the critical configurations may alter for various choices of the sink, whereas the spanning trees of G are fixed regardless of the sink. To overcome this difficulty, we define the *level* of a configuration.

Definition 3.2.1. For a configuration $\mathbf{c}$ on a graph $G = (V, E)$ with sink vertex q, the *level* of $\mathbf{c}$ is

$$\text{level}(\mathbf{c}) = \text{wt}(\mathbf{c}) - |E| + \deg(q).$$

Example 3.2.2. Returning to the example $G = K_4 \backslash e$, for the first choice of sink, G_1 of Figure 3.2, the total number of edges is 5 and the degree of the sink is 2. For the second choice of sink, G_2 of Figure 3.2, the total number of edges remains 5 but the degree of the sink is 3. The critical configurations of these two graphs and their levels are displayed below.

G_1		G_2	
criticals	level	criticals	level
$(1, 2, 2)$	2	$(1, 2, 1)$	2
$(0, 2, 2)$	1	$(1, 1, 1)$	1
$(1, 1, 2)$	1	$(0, 2, 1)$	1
$(1, 2, 1)$	1	$(1, 2, 0)$	1
$(0, 2, 1)$	0	$(0, 2, 0)$	0
$(1, 0, 2)$	0	$(0, 1, 1)$	0
$(0, 1, 2)$	0	$(1, 1, 0)$	0
$(1, 2, 0)$	0	$(1, 0, 1)$	0

The critical configurations themselves have changed but the multiset of levels of the critical configurations are the same for both choices of the sink vertex.

We will see below in Corollary 3.3.2 that for any graph, the multiset of levels is the same for any choice of sink vertex. Thus the level gives a statistic on configurations involving the number of chips in the configuration that is invariant to the choice of sink vertex.

Proposition 3.2.3. *Let $\mathbf{c}$ be a critical configuration for a graph $G = (V, E)$ with n vertices, then*

$$0 \leq \text{level}(\mathbf{c}) \leq |E| - n + 1.$$

Proof. One direction is established by bounding the level of any critical configuration by the level of $\mathbf{c}_{\max}$. Recall that for a finite graph G with choice of sink q, there is a unique maximal critical configuration $\mathbf{c}_{\max}$ equal to the degree sequence of G minus one in each coordinate. The configuration $\mathbf{c}_{\max}$ is coordinate-wise larger than all other critical configurations and hence has the maximal level of all critical configurations. For any configuration $\mathbf{c}$,

$$
\begin{aligned}
\text{level}(\mathbf{c}) \ &\leq\ \text{level}(\mathbf{c}_{\max}) \\
&=\ \sum_{v \neq q} (\deg(v) - 1) - |E| + \deg(q) \\
&=\ \sum_{v \neq q} (\deg(v) - 1) + \deg(q) - |E| \\
&=\ 2|E| - (n - 1) - |E| \\
&=\ |E| - n + 1.
\end{aligned}
$$

For the other direction, consider a critical configuration $\mathbf{c}$. Condition 8 of Theorem 2.6.3 gives that $\mathbf{c}$ is stable and that after firing the sink vertex, every non-sink vertex fires exactly once before restabilizing. Consider firing the sink vertex from $\mathbf{c}$. As in the proof of Theorem 2.3.6 part (c), associate to each edge the first chip to cross the edge. When the configuration restabilizes, $\deg(q)$ many chips are back at the sink vertex and $|E| - \deg(q)$ are on non-sink vertices, therefore

$$
\sum_{v \neq q} c_v \geq |E| - \deg(q).
$$

$\square$

Note that dually, this gives that the total number of chips of any superstable configuration is bounded above by $|E| - (n - 1)$.

Definition 3.2.4. The *critical polynomial* of a graph G is the generating function for critical configurations by level.

$$
P_G(y) = \sum_{\mathbf{c} \in \text{Crit}(G)} y^{\text{level}(\mathbf{c})},
$$

where $\text{Crit}(G)$ is the collection of all critical configurations of G for any fixed choice of sink vertex.

Again, we will see shortly that the critical polynomial is well-defined for a graph G because the multiset of levels is the same for any choice of sink.

The bound on level given in Proposition 3.2.3 can be interpreted as a bound on the degree of the critical polynomial.

3.2.2   Activity

We have now seen that critical configurations can be graded by level. The corresponding statistic for spanning trees is the external activity of the tree. External activity is a well-studied matroid property. We take a moment to consider the collection of spanning trees of a graph as the collection of bases of a matroid.

Although matroid theory is the appropriate setting for the following results, it is not strictly necessary and we include an explanation of all results at the level of graph theory. A good reference for the matroid theory discussed here is Oxley's text [Oxl11].

Definition 3.2.5 (Graphical Matroid 1). For a finite graph $G = (V, E)$, the *graphical matroid* $M(G)$ is the matroid with ground set equal to E and independence given by:

1. A collection of edges $F \subseteq E$ is independent if it is acyclic.

2. A collection of edges $T \subseteq E$ is a basis if it forms a spanning tree.

Equivalently, the graphical matroid can be defined via the oriented incidence matrix.

Definition 3.2.6 (Graphical Matroid 2). For a finite graph $G = (V, E)$, the *graphical matroid* $M(G)$ is the vector matroid given by the column vectors of the oriented incidence matrix ∂ corresponding to any choice of ordering of the edges of G.

Example 3.2.7. Consider the edge labeled $K_4 \setminus e$ of Figure 3.3.

FIGURE 3.3: Bases of the graphical matroid correspond to spanning trees or equivalently to column bases of the oriented incidence matrix.

Suppose that all edges are oriented towards the sink. The oriented incidence matrix for this graph is:

$$
\partial_{K_4 \setminus e} =
\begin{array}{c}
v_1 \\ v_2 \\ v_3 \\ q
\end{array}
\begin{array}{cccccc}
1 & 2 & 3 & 4 & 5 \\
\left(\begin{array}{ccccc}
0 & 1 & -1 & 0 & 0 \\
0 & 0 & 1 & 1 & 1 \\
1 & 0 & 0 & -1 & 0 \\
-1 & -1 & 0 & 0 & -1
\end{array}\right)
\end{array}.
$$

The collection of edges $\{2, 4, 5\}$ is a basis of $M(G)$ because the corresponding edges form a spanning tree. Equivalently the corresponding columns of $\partial_{K_4 \backslash e}$ form a column basis of $\partial_{K_4 \backslash e}$.

3.2.3 The Tutte polynomial

The circuits of a graphical matroid $M(G)$ are the collections of edges of the circuits of G.

Definition 3.2.8. The *fundamental circuit of a spanning tree T with respect to e* is the unique cycle formed when an edge $e \notin T$ is added to T.

Dually, the cocircuits of a graphical matroid $M(G)$ are minimal edge cuts of G.

Definition 3.2.9. The *fundamental cocircuit of a spanning tree T with respect to $e \in T$* is the unique edge cut disjoint from $T \backslash e$.

An edge e is *externally active with respect to a tree T* if e is the edge with smallest label in the fundamental circuit of T. Similarly, an edge e is *internally active with respect to a tree T* if e is the edge with smallest label in the fundamental cocircuit of T.

We write $ia(T)$ for the number of internally active elements in a tree T and $ea(T)$ for the number of externally active elements in a tree T.

Example 3.2.10. Consider again $G = K_4 \backslash e$ labeled as in Figure 3.3.

- Let T be the spanning tree (basis) $\{2, 4, 5\}$.

- For $e = 1$, the fundamental circuit $C(T, e)$ is the cycle $\{1, 4, 5\}$. Since $e = 1$ is the smallest element of $C(T, e)$, e **is** externally active with respect to T.

- For $e = 4$, the fundamental cocircuit $C_o(T, e)$ is the cut $\{1, 4\}$. Since $e = 4$ is not the smallest element of $C_o(T, e)$, e **is not** internally active with respect to T.

The Tutte polynomial records the activities of trees (bases) in a generating function.

Definition 3.2.11 (Tutte polynomial 1). Let G be a graph with collection of spanning trees $\mathcal{T}$ (equiv. let G be a graphical matroid with collection of bases $\mathcal{T}$). The *Tutte polynomial* of G is the two-variable generating function

$$T_G(x, y) = \sum_{T \in \mathcal{T}} x^{ia(T)} y^{ea(T)}.$$

We may be concerned with the choice of ordering of the ground set in the definitions of activity. This choice however does not change the Tutte polynomial – similar to the choice of sink with respect to the number of critical configurations or the labeling of edges for the construction of the oriented incidence matrix. The Tutte polynomial has several alternative characterizations. Tutte's original formulation does not make use of a choice of labeling. It is formulated in terms of the rank function.

For a graphical matroid, with ground set E, a subset A of E is a collection of edges. The *matroid rank* $\mathrm{rk}(A)$ is equal to the size of the largest forest (independent set) contained in A.

Definition 3.2.12 (Tutte polynomial 2). The *Tutte polynomial* of G is equal to

$$T_G(x, y) = \sum_{A \subseteq E} (x - 1)^{\mathrm{rk}(E) - \mathrm{rk}(A)} (y - 1)^{|A| - \mathrm{rk}(A)}.$$

The Tutte polynomial carries much information about the matroid. For the graphical matroid $M(G)$ of a graph G on n vertices, evaluations of the Tutte polynomial yield:

- $T_{M(G)}(1, 1)$ equals the number of spanning trees of G.

- $T_{M(G)}(2, 0)$ equals the number of acyclic orientations of G.

- $T_{M(G)}(0, 2)$ equals the number of totally cyclic orientations of G.

- $T_{M(G)}(1, 0)$ equals the number of acyclic orientation of G with a unique fixed source.

- $T_{M(G)}(1 - x, 0)t^{\kappa(G)}$ is the chromatic polynomial of G.

- $T_{M(G)}(1, y)$ gives the generating function for spanning trees ranked by external activity.

Example 3.2.13. Continuing the example above, the Tutte polynomial of $K_4 \backslash e$ is:

$$T_{K_4 \backslash e} = x^3 + 2x^2 + x + 2xy + y + y^2.$$

The following two evaluations are consistent with our earlier calculations

$$T(1, 1) = 8$$
$$T(1, y) = y^2 + 3y + 4.$$

The graph $K_4 \setminus e$ has 8 spanning trees:

1 with external activity equal to 2,
3 with external activity equal to 1,
4 with external activity equal to 0.

An edge is a *coloop* of a graph if it is contained in every spanning tree.

Proposition 3.2.14. *The Tutte polynomial of any matroid satisfies a deletion contraction recursion. For a graph G and edge e,*

$$T_G = T_{G-e} + T_{G/e}, \ \text{if } e \text{ is not a loop or coloop}$$
$$T_G = T_e T_{G-e}, \ \text{if } e \text{ is a loop or coloop}$$

with base cases, $T_{coloop} = x$, *and* $T_{loop} = y$.

The recursive formulation of the Tutte polynomial is particularly conducive to computations. There is also a beautiful universality result of recursive invariants.

A *Tutte–Grothendieck (or contraction-deletion)* invariant of a graph G is an invariant $f(G)$ that satisfies:

$$f_G = f_{G-e} + f_{G/e}, \ \text{if e is nontrivial}$$
$$f_G = f_e f_{G-e}, \ \text{if e is a loop or coloop.}$$

The Tutte polynomial is the universal deletion–contraction invariant in that *any* contraction-deletion invariant of a graph is an evaluation of the Tutte polynomial. Below, I refers to a single isthmus (a graph with two vertices and a single edge) and J refers to a single loop.

Theorem 3.2.15. *The Tutte polynomial is the unique Tutte–Grothendieck isomorphism invariant from the class of matroids into the polynomial ring* $\mathbb{Z}[x,y]$ *satisfying*

$$T(I; x, y) = x$$
$$T(L; x, y) = y.$$

We have presented only a narrow discussion of the Tutte polynomial. We refer the reader to [EMM11] for a comprehensive survey of the Tutte polynomial and its applications.

3.3 Merino's theorem

We are now ready to state Merino's Theorem. The result was first conjectured by Biggs and appears in [Mer01].

Theorem 3.3.1 (Merino's Theorem). *For a graph G, for any choice of vertex, the critical polynomial is equal to the Tutte polynomial evaluated at $(1, y)$:*

$$T_G(1, y) = P_G(y).$$

Proof. The proof establishes the deletion–contraction relation for the critical polynomial and then proceeds by induction.

We will sketch the various cases leaving the details as an exercise.

Throughout the proof, let $G = (V, E)$ and fix an edge $e = (u, q)$ which is incident to the sink vertex q.

- Suppose e is the only edge of G. If e is a coloop (G has two vertices and a single edge) then the only critical configuration is the all zeros configuration and

$$P_G(y) = 1.$$

  If e is a loop (G has a single vertex and a single edge) then the only critical configuration is a single zero and

$$P_G(y) = y.$$

- Suppose $|E| \geq 2$ and e is a coloop. The critical configurations of G correspond to critical configurations of G/e as follows. Suppose that $\mathbf{c}$ is a critical configuration for G/e then define the configuration $\mathbf{c}'$ on G to be equal to $\mathbf{c}$ on all vertices $v \neq u, q$. Set the value at u to $\deg(u) - 1$. Note that when e is a coloop, in any critical configuration $\mathbf{c}$ on G, the value of $\mathbf{c}_u$ must be $\deg(u) - 1$. It can then be shown that $\text{level}(\mathbf{c}) = \text{level}(\mathbf{c}')$ and

$$P_G(y) = P(G/e, y).$$

- Suppose $|E| \geq 2$ and e is a loop. The critical configurations of G correspond to critical configurations of $G - e$. If $\mathbf{c}$ is a critical configuration of G then it is also a critical configuration for $G - e$. The level of the configuration changes by 1 and

$$P_G(y) = yP_{G-e}(y).$$

- For the generic case, suppose that $|E| \geq 2$ and e is neither a loop nor coloop. This situation is broken into two cases which follow the same ideas as the cases above. Those critical configurations $\mathbf{c}$ of G with $\mathbf{c}_u = \deg(u) - 1$ will correspond to critical configurations of the contraction G/e. Those critical configurations that do not have a maximal value at the vertex u will correspond to critical configurations of $G-e$. In both cases, the level of the configurations do not change. Therefore we can conclude

$$P_G(y) = P_{G-e}(y) + P_{G/e}(y).$$

□

Corollary 3.3.2. *For any graph, the multiset of levels is the same for any choice of sink vertex.*

Proof. The Tutte polynomial does not depend on a choice of sink vertex.

□

Example 3.3.3. In this example we illustrate two instances from the proof of Merino's Theorem for the graph $K_4 \setminus e$.

Suppose $u = v_2$ and $e = \{v_2, q\}$.

Consider the critical configuration: $\mathbf{c} = (1, 0, 2)$. In this configuration, $\mathbf{c}_u \neq \deg(u) - 1$. We can check that $(1, 0, 2)$ is also a critical configuration for $G - e$ by firing the sink vertex and confirming that the configuration returns to $(1, 0, 2)$.

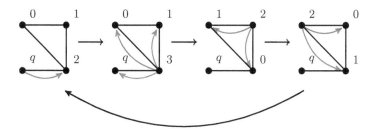

Furthermore we check that the levels remain the same:

$$\text{level}_G(\mathbf{c}) = 3 - 5 + 2 = 0$$
$$\text{level}_{G-e}(\mathbf{c}) = 3 - 4 + 1 = 0.$$

Next consider the critical configuration: $\mathbf{c} = (1, 2, 2)$. In this configuration, $\mathbf{c}_u = \deg(u) - 1$. We can check that $(1, 2)$ is a critical configuration for G/e by again firing the sink vertex and confirming that the configuration returns to $(1, 2)$.

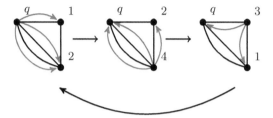

Furthermore we check that the levels remain the same:

$$\text{level}_G(\mathbf{c}) = 5 - 5 + 2 = 2$$
$$\text{level}_{G/e}(\mathbf{c}) = 3 - 4 + 3 = 2.$$

From above, we know that $T_{K_4 \setminus e}(1, y) = y^2 + 3y + 4$. And indeed there are four critical configurations with level 0, three critical configurations with level 1 and a unique critical configuration with level 2 for this graph.

3.3.1 The O-conjecture

This section takes a detour into the combinatorics of simplicial complexes to show an important application of the theory of chip-firing to algebraic combinatorics. The next section continues the narrative of bijections between critical configurations and spanning trees and does not depend on the material here.

For any simplicial complex Σ, an important combinatorial invariant is the number of simplicies of each dimension, the face numbers of Σ. The *f-vector* of Σ records this information as an integer sequence:

$$f(\Sigma) = (f_{-1}, f_0, f_1, \ldots, f_d),$$

where f_i = the number of faces of dimension i. A graph may be considered as a one-dimensional cell complex. The edges are the one-dimensional faces, the vertices are the zero-dimensional faces, and the empty set is the unique face recorded as dimension -1; see Section 7.2 for an introduction to simplicial complexes.

For a complex Σ of dimension d, the generating function of face numbers is

$$f_\Sigma(t) = \sum_{i=0}^{d} f_i t^{d-i}.$$

Much work in algebraic combinatorics has focused on face numbers both in understanding which vectors can appear for various classes of complexes and the structure of these sequences.

Often it is more natural to work with a transformation of this data, as will be the case here.

The *h-vector* of a complex Σ,

$$h(\Sigma) = (h_0, h_1, \ldots, h_{d+1}),$$

is defined in terms of the f-vector as follows

$$h_k = \sum_{i=0}^{k} (-1)^{k-i} \binom{d+1-i}{k-i} f_{i-1}, \qquad 0 \le k \le d+1.$$

The generating function of the h-numbers is

$$h_\Sigma(t) = \sum_{i=0}^{d} h_i t^{d-i}.$$

As polynomials the relationship between f and h simplifies to

$$h_\Sigma(t+1) = f_\Sigma(t).$$

For a matroid M, the simplicial complex of interest is the *independent set complex* $\Sigma(M)$ of M. This is a simplicial complex with vertex set equal to the ground set of M. A subset $I \subseteq [E]$ is in $\Sigma(M)$ if I is an independent set of M.

A complex is *pure* if all maximal faces have the same dimension. Matroid independent set complexes are always pure but can be characterized in terms of purity.

Example 3.3.4. The independent set complex of $M(K_4 \setminus e)$ is a two-dimensional simplicial complex on 5 vertices corresponding to the 5 edges of $K_4 \setminus e$.

There are 8 two-dimensional faces. The independent set complex consists of these 8 faces and all subsets. The bracket notation denotes "generated by."

$$\Sigma(M(K_4 \setminus e)) = \langle 123, 145, 234, 235, 245, 345, 124, 135 \rangle,$$

$$f_\Sigma = (1, 5, 10, 8),$$
$$h_\Sigma = (1, 2, 3, 2).$$

The f-vector of the independent set complex of a matroid records the number of independent sets of each size. The h-vector is a transformation of this data and has very nice interpretations itself. The h-vector of a matroid complex is always positive. It is known to enumerate bases by internal activity. Namely, the h-vector is the sequence of coefficients of the Tutte polynomial evaluated at $(x, 1)$:

$$h_{\Sigma(M)}(t) = T_M(x, 1).$$

Example 3.3.5. From Example 3.2.13, the Tutte polynomial of $M(K_4 \backslash e)$ is:

$$T_{K_4 \backslash e} = x^3 + 2x^2 + x + 2xy + y + y^2.$$

Evaluating at $(x, 1)$ gives

$$T_{K_4 \backslash e}(x, 1) = x^3 + 2x^2 + 3x + 2,$$

which matches the h-vector calculation above.

Matroid duality allows us to recover the evaluation $(1, y)$ that appeared in the previous section. Let M be a matroid and M^* its dual matroid. The Tutte polynomial behaves nicely under duality,

$$T_M(x, y) = T_{M^*}(y, x).$$

A matroid M is called *cographic* if the dual matroid M^* is graphical, i.e. it is the graphical matroid for some graph G.

Combining Merino's Theorem and the duality relation for the Tutte polynomial, gives the following.

Proposition 3.3.6. *Suppose M is a cographic matroid and M^* is the graphical matroid of a graph G then*

$$h_{\Sigma(M)}(t) = T_M(x, 1) = T_{M^*}(1, y) = P_G(y).$$

Stanley's O-conjecture is about the structure of the h-vectors of matroids.

Consider a collection of integer vectors of $\mathbb{Z}^n$ under the *componentwise partial order*: For $\mathbf{a} = (a_1, a_2, \ldots, a_n) \in \mathbb{Z}^n$ and $\mathbf{b} = (b_1, b_2, \ldots, b_n) \in \mathbb{Z}^n$, $\mathbf{a} \leq \mathbf{b}$ if $a_i \leq b_i$ for all i. An *order ideal* I in this partial order is any collection of integer vectors such that if $\mathbf{b} \in I$

and $\mathbf{a} \leq \mathbf{b}$ then $\mathbf{a} \in I$. Namely, the collection is downward closed in the componentwise partial order.

The *degree* of an integer sequence is the sum of its entries. (These definitions are often given in terms of monomial ideals of a polynomial ring – a collection is an order ideal if the monomials are closed under division and the degree is the standard sum of exponents).

An *O-sequence* is any integer sequence which can be interpreted as counting the number of elements (monomials) of each degree in an order (monomial) ideal.[1] Finally, a *pure O-sequence* is one arising from a pure order (monomial) ideal, i.e. all maximal elements (monomials) have the same degree.

Conjecture 3.3.7 (Stanley's O-conjecture, see [Sta96a]). *For all matroids M, the h-vector of $\Sigma(M)$ is a pure O-sequence.*

The conjecture has been verified for all matroids on ground sets of size at most 9 [DLKK12]. It is also known to hold for a number of special cases, for example for lattice path matroids [Sch10], co-transversal matroids [Oh13], paving matroids [MNRIVF12] and cographic matroids [Mer01]. The proof for cographic matroids follows from Merino's Theorem.

Theorem 3.3.8 ([Mer01]). *The O-conjecture holds for cographic matroids.*

Proof. The evaluation of the Tutte polynomial considered in Merino's Theorem is $T_{M(G)}(1, y)$ which gives the h-vector of the dual matroid $M^*(G)$. These coefficients also give the number of superstable configurations of each level. By Proposition 2.6.22 of Chapter 2, the superstable configurations of G form a pure order ideal. $\square$

Example 3.3.9. Let $M(K_4 \backslash e)$ be the graphical matroid of the graph $K_4 \backslash e$ as labeled in Example 3.3.4. The dual matroid M^* is also a graphical matroid. It is the graphical matroid of the graph consisting of a triangle with two double edges.

The bases of M^* are $\{25, 15, 45, 35, 12, 34, 13, 24\}$. The complex $\Sigma(M^*)$ is a graph on 5 vertices. The face numbers are given by:

$$f_{M^*} = (1, 5, 8)$$
$$h_{M^*} = (1, 3, 4).$$

Recall that $T_{K_4 \backslash e}(1, y) = y^2 + 3y + 4$.

---

[1] In some references, O-sequences are called M-sequences.

Up to isomorphism, there are two choices for a sink in $K_4 \backslash e$. Each choice gives a different collection of superstable configurations; see Example 3.1.4. Either one gives a pure monomial ideal with the number of monomials of each degree given by the entries of h_{M^*} as shown in Figure 3.4.

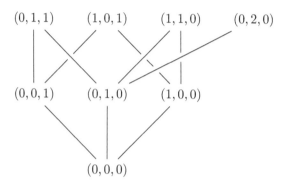

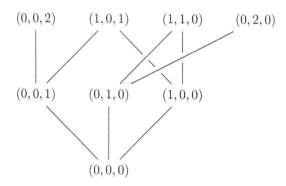

FIGURE 3.4: The two possible posets of superstable configurations for $K_4 \setminus e$ for the two possible choices of sink vertex.

Thus superstable configurations of the chip-firing process form a pure order ideal in the componentwise partial order with the number of elements of each degree given by activity. And thus superstable configurations provide witness for cographic matroids for Stanley's conjecture.

3.4   Cori–Le Borgne bijection

In this section we return to bijections between spanning trees and superstable configurations. The Cori–Le Borgne bijection is an activity preserving bijection between spanning trees and superstable configurations. It is constructed via a modification of Dhar's burning algorithm.

Recall that the burning algorithm is a method for checking whether or not a chip configuration is superstable. A fire is started at the root vertex and is able to spread across edges of the graph to burn other vertices. The chips of a configuration represent firefighters. The firefighters at a vertex can each turn and protect the vertex from an approaching fire along a single edge. If, however, at any time there are more burning incident edges than firefighters at a vertex then the vertex itself burns and the fire spreads along incident edges. A chip configuration is superstable if and only if every vertex is eventually burnt in this process.

The modification we now add is to burn edges one at a time. First, the edges of the graph are (arbitrarily) ordered. A fire is started at the sink vertex. The fire then burns along the edge incident to the sink of largest label. At each time step, if the fire can burn a vertex it does so. Also, at each time step, only a single edge is burnt – the edge of largest label that is susceptible to the fire. When a fire burns through a vertex v there is then a well-defined last edge that was burnt before burning through v. For each vertex, mark this last edge.

Starting with a superstable configuration, all vertices will eventually be burnt – this is Dhar's algorithm. We furthermore now claim that the collection of marked edges forms a spanning tree whose external activity is equal to the level of the configuration.

Cori–Le Borgne burning algorithm:

1. Input a superstable configuration.

2. Fix an ordering on the edges of G.

3. Start a fire at the sink vertex q and follow the rules of Dhar's algorithm except:

4. If at any step, the fire can burn through multiple edges, burn only through the largest edge.

5. When a fire reaches a vertex, mark the last edge burnt before going through the vertex.

6. Output the collection of marked edges.

For a superstable configuration **s**, let $\Phi(\mathbf{s})$ be the output of the Cori–Le Borgne algorithm.

Theorem 3.4.1 ([CLB03]). *Let G be a graph on n vertices and* **s** *a superstable configuration of G.*

1. *The collection of marked edges $\Phi(\mathbf{s})$ forms a spanning tree.*

2. *The map Φ from superstable configurations to spanning trees is a bijection.*

3. *If $\Phi(\mathbf{s}) = T$, then*

$$\mathrm{level}(\mathbf{c}_{\max} - \mathbf{s}) = ea(T),$$

   the external activity of T.

Proof. (sketch)
For item 1, the input **s** is a superstable configuration; by Dhar's original burning algorithm, the fire will eventually burn through all vertices. Therefore $\Phi(\mathbf{s})$ will have cardinality $n-1$. By construction, the collection of edges is connected. Together these imply $\Phi(\mathbf{s})$ must be a spanning tree.

For item 2, we describe the inverse map associating a superstable configuration to every tree. Given a spanning tree T, generate a chip configuration **c** as follows. Start a fire at the sink vertex q. Edges are burnt as above. If at any step, the fire can burn through multiple edges, burn only through the smallest edge. At each vertex, the fire is stopped along edges not in T. When a fire burns through a vertex v, set $\mathbf{c}_v$ equal to the number of burnt edges incident to v minus 1.

For item 3, we do not go through the technical aspects of the bijection but do provide an example below. □

Example 3.4.2. Figure 3.5 shows an instance of the Cori–Le Borgne algorithm.

- The fire starts along edge 5 and is blocked.

- The fire starts again from q and burns through the edge 3.

- At the next stage of the algorithm, both edges 1 and 4 are eligible to burn. Edge 4 burns because it has a larger label but is blocked before reaching the next vertex.

- At the next stage only edge 1 is eligible to burn, hence edge 1 is the next to burn even though edge 2 has larger label. This fire is not blocked and burns through to edge 2.

- The collection of marked edges is $\{1,2,3\}$.

We confirm that the level of the input equals the activity of the output. First, find the critical configuration dual to the input superstable configuration and then compute its level:

$$\mathbf{c}_{\max} - (0,1,1) = (1,1,0)$$

$$\text{level}(1,1,0) = 2 - 5 + 3 = 0.$$

Second, note that neither edge 4 nor edge 5 is the smallest labeled edge in the fundamental circuit created by adding the edge. Therefore the external activity of the tree $\{1,2,3\}$ is 0.

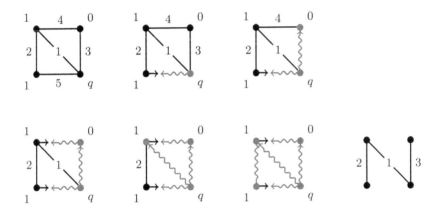

FIGURE 3.5: An example of the extended burning algorithm used in the Cori–Le Borgne bijection.

3.5 Acyclic orientations

We have now established strong connections between the long-term stable chip configurations of the chip-firing process and an important combinatorial structure – spanning trees. In the next few sections, we further refine the association and make connections to other important graphical structures, starting with acyclic orientations and graphical hyperplane arrangements.

Recall that *maximum* superstable configurations are those configurations with the largest weight (sum of the number of chips on non-sink vertices). *Maximal* superstable configurations are those superstable configurations **c** such that no other superstable configuration **b** is larger in the componentwise partial order, i.e. **c** is maximal if there does not exist a superstable configuration **b** such that **b** $\geq$ **c**.

Proposition 2.6.22 shows that for finite undirected graphs, maximum and maximal configurations coincide. Hence all componentwise largest superstable configurations have the same number of chips.

The following theorem puts maximal superstable configurations in bijective correspondence with certain acyclic orientations. The result appears as [BCT10, Theorem 3.1] and [Big99a, Lemma 10], see also [GZ83]. The connection will play an important role in the Riemann–Roch Theorem of Chapter 8.

Theorem 3.5.1. *Let G be a graph on n vertices with sink q. Let $AO(G)_q$ be the collection of acyclic orientations of G with unique sink q. Let $\mathrm{maxSS}(G)_q$ be the collection of maximal superstable configurations of G with sink q. Then the map*

$$f : AO(G)_q \to \mathrm{maxSS}(G)_q$$
$$f(\mathcal{O}) = \mathrm{outdeg}(\mathcal{O}) - \mathbf{1}$$

is a bijection.

Proof. Take an acyclic orientation $\mathcal{O}$ with unique sink at q. Let $\mathbf{c}_\mathcal{O}$ be the chip configuration $\mathrm{outdeg}(\mathcal{O}) - \mathbf{1}$ as above.

For any acyclic orientation, the sum of the outdegrees is equal to the number of edges. Therefore

$$\sum_v \mathbf{c}_v = |E| - (n-1)$$

which is the maximum value of a superstable configuration (as noted above after the bound on level).

The configuration $\mathbf{c}_\mathcal{O}$ is superstable; this can be seen via Dhar's algorithm. If we momentarily disregard the sink vertex, the resulting configuration and corresponding acyclic orientation is the same as in Theorem 2.3.6. At the first stage of Dhar's algorithm, any sink vertices of the acyclic orientation of $V \setminus q$ will be burnt. As the fire burns through, neighboring vertices will devote one chip to stopping the fire, again akin to the proof of Theorem 2.3.6 where we consider reversing the orientation of each edge incident to a source vertex. The new configuration also corresponds to an acyclic orientation and the process continues until all vertices are burnt.

The correspondence is injective because acyclic orientations are uniquely determined by their outdegree sequences.

Finally, the correspondence is surjective. Given a maximal super-stable configuration **s**, we construct an acyclic orientation $\mathcal{O}$ such that $\mathbf{s} = \text{outdeg}(\mathcal{O}) - \mathbf{1}$. First, orient all edges incident to q towards q. At any step, if the outdegree of any vertex v reaches the value $s_v + 1$ then orient all remaining unoriented edges incident to v towards v.

$\square$

Combining Theorem 3.5.1 with the results of the previous sections gives the following list of equivalences.

Theorem 3.5.2. *For a graph G with sink q, the following are the same:*

1. *The number of maximal superstable configurations of G.*

2. *The number of acyclic orientations of G with unique sink q.*

3. *The number of spanning trees of G with zero external activity (i.e. with no broken circuits).*

4. *The coefficient of the linear term of the chromatic polynomial of G.*

5. *The evaluation of the Tutte polynomial $T(1,0)$.*

Theorem 3.5.2 brings together a number of results from slightly different perspectives. References are well listed in [BCT10] where this is Theorem 4.1.

Example 3.5.3. Let $G = K_4 \backslash e$ with sink q as in Example 3.4.2.

- G has 4 maximum (= maximal) superstable configurations, as seen in Example 3.3.9.

- Consider orienting the three edges incident to q towards q. Two edges remain. Any combination of orientations for the remaining two edges yields an acyclic orientation with unique sink q, giving 4 in total. The orientations and corresponding superstable configurations appear in Figure 3.6.

- From Example 3.2.13 we have that $T(1, y) = y^2 + 3y + 4$. And we confirm that $T(1, 0) = 4$.

3.5.1 Hyperplane arrangements

In this section, we connect chip-firing to graphical hyperplane arrangements. For background on the combinatorics of hyperplane arrangements see [Sta07].

Associated to any graph is a graphical hyperplane arrangement.

Definition 3.5.4. Let $G = (V, E)$ be a finite undirected graph on n vertices. The *graphical arrangement* of G is the hyperplane arrangement $\mathcal{A}_G$ in $\mathbb{R}^n$ consisting of the hyperplanes

$$\mathcal{A}_G = \{x_i = x_j \mid \{i, j\} \in E\}.$$

For the case of the complete graph, $G = K_n$, the graphical arrangement is the well-known *Braid arrangement* $\mathcal{B}_n$ consisting of all hyperplanes $x_i = x_j$ for all $i, j \in [n]$.

Proposition 3.5.5. *The regions of the graphical arrangement $\mathcal{A}_G$ are in bijection with the acyclic orientations of G.*

Proof. Consider a point in the interior of a region of $\mathcal{A}_G$. For each hyperplane, record on which side of the hyperplane it lies, either $x_i > x_j$ or $x_i < x_j$. Construct an orientation of G by orienting $i \to j$ if $x_i > x_j$ and $j \to i$ if $x_i < x_j$. In this way, each edge receives an orientation. The orientation is acyclic because otherwise we would have $z_{i_1} > z_{i_2} > \cdots > z_{i_k} > z_{i_1}$ for some point $z \in \mathbb{R}^n$. $\qquad\square$

An important theorem in the combinatorial theory of hyperplane arrangements is the Greene–Zaslavsky Theorem stating that the number of regions of a hyperplane arrangement is a simple evaluation of the characteristic polynomial of the arrangement. We rephrase it here in terms of the matroid associated with the arrangement. In general, for a finite real linear arrangement $\mathcal{A}$, the matroid $M(\mathcal{A})$ is the vector matroid of the collection of all normal vectors to all hyperplanes in $\mathcal{A}$.

Theorem 3.5.6 ([GZ83]). *The number of regions of a finite real hyperplane arrangement $\mathcal{A}$ is an evaluation of the characteristic polynomial of $\mathcal{A}$ which in turn is an evaluation of the Tutte polynomial:*
The number of regions of $\mathcal{A} = T_{M(\mathcal{A})}(2, 0)$.

For a graph $G = (V, E)$ and a vertex $q \notin V$, define $G * q$ to be the graph with vertex set $V \cup q$, all edges of G, and an edge between q and all vertices in V.

As a corollary of the Greene–Zaslavsky Theorem, we extend Theorem 3.5.1 to include:

Theorem 3.5.7. *For a graph $G = (V, E)$ and $q \notin V$, the following are the same:*

1. *The number of regions of $\mathcal{A}_G$.*

2. *The number of acyclic orientations of G.*

3. *The number of acyclic orientations of $G * q$ with unique sink q.*

4. *The number of maximal superstables of $G * q$ with sink q.*

5. *The evaluation $T_{M(G)}(2, 0)$ of the Tutte polynomial.*

Example 3.5.8. Let G be the two-edge path consisting of edges $\{12\}$ and $\{13\}$. The arrangement $\mathcal{A}_G$ consists of two hyperplanes in $\mathbb{R}^3$. Figure 3.6 shows a two-dimensional representation which captures the combinatorics of the arrangement. The first diagram shows G along with the coordinate constraints induced by $\mathcal{A}_G$. The second diagram shows $G * q$ with q as the unique sink. The outdegree sequences of the induced acyclic orientations give the superstable configurations of $G * q$.

In the next section we consider a larger hyperplane arrangement which contains the graphical arrangement as a subarrangement. This larger arrangement captures all superstable configurations, not necessarily bijectively but at least surjectively.

3.6 Parking functions

Parking functions are another well-studied combinatorial object closely related to spanning trees.

Definition 3.6.1. A *parking function* $p = (p_1, p_2, \ldots, p_n) \in \mathbb{Z}^n$ is an integer sequence satisfying the following criterion:
   If $q_1 \leq q_2 \leq \cdots \leq q_n$ is the non-decreasing rearrangement of p, then

$$q_i \leq i \text{ for all } i \in [n].$$

See Exercise 3.9.11 for a description of parking functions as preference orders for parking cars, hence motivating the name.

The number of parking functions of length n is $(n + 1)^{n-1}$ as first shown in [Pyk59], and [KW66]. The reader may also recognize this quantity as the number of spanning trees of the complete graph K_n.

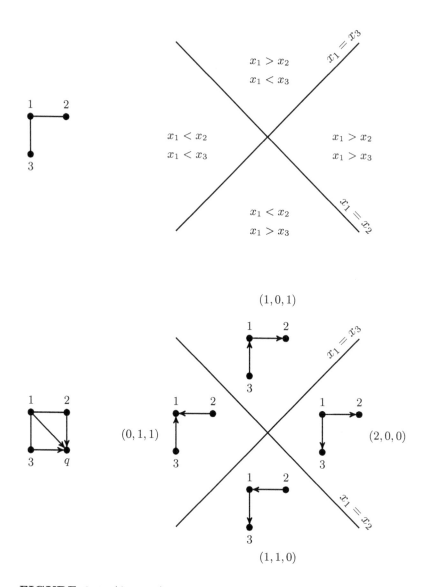

FIGURE 3.6: Above: An example of a graphical arrangement for a graph G and the induced acyclic orientations by coordinate constraints. Below: The acyclic orientations of G and induced superstable configurations of $G * q$ with q as the unique sink.

For a graph $G = (V, E)$, a subset of vertices $A \subseteq V$, and a vertex $v \in A$, define the *outdegree of v with respect to A* as the number of vertices not in A that are incident to v:

$$\mathrm{outdeg}_A(v) = |\{u \ : \ \{u, v\} \in E, \ u \notin A\}|.$$

Definition 3.6.2. Let G be a graph with sink q. A *G-parking function* is a function $p : V(G) \setminus q \to \mathbb{Z}$ such that for every non-empty subset $A \subseteq V \setminus q$, there exists a vertex $v \in A$ such that

$$p(v) < \mathrm{outdeg}_A(v).$$

K_{n+1}-parking functions are precisely the parking functions of Definition 3.6.1. The following theorem connects parking functions and chip-firing.

Theorem 3.6.3. *Let G be a graph with sink vertex q. Then the G-parking functions of G are precisely the set of superstable configurations of G.*

Proof. This is simply a matter of reinterpreting the definition above. Let p be a G-parking function of length n. Then for any subset A of non-sink vertices, there will be some vertex in G which makes the cluster-fire of vertices in A illegal. $\square$

Returning to the case $G = K_n$, Pak and Stanley gave a bijective correspondence between parking functions (superstable configurations) and regions of the Shi hyperplane arrangement.

Definition 3.6.4. The *Shi arrangement* $\mathcal{A}_{\mathrm{shi}(n)}$ is the hyperplane arrangement in $\mathbb{R}^n$ consisting of the hyperplanes

$$\{x_i = x_j, \quad x_i = x_j + 1\} \text{ for all } i < j \leq n.$$

The *central region* of the Shi arrangement is the region such that

$$0 < x_i - x_j < 1.$$

It was well known that the number of regions of the Shi arrangement is $(n + 1)^{n-1}$. Pak and Stanley gave an explicit labeling of the regions of the arrangement by parking functions:

Pak–Stanley labeling

1. Input $\mathcal{A}_{\mathrm{shi}(n)}$.

2. Label the central region $(0, 0, \ldots, 0)$.

3. Starting from the central region and walking to adjacent regions label regions by the following rules:

4. If a hyperplane of the form $x_i - x_j = 0$ is crossed, increase the jth coordinate of the label by 1.

5. If a hyperplane of the form $x_i - x_j = 1$ is crossed, increase the ith coordinate by 1.

Example 3.6.5. Figure 3.7 shows the Shi arrangement $\mathcal{A}_{\text{shi}(3)}$ and the Pak–Stanley labels. The labels are precisely the superstable chip configurations corresponding to the complete graph on 4 vertices $K_4 = K_3 * q$.

Theorem 3.6.6 ([Sta96b]). *The region labels generated by the Pak–Stanley labeling of the Shi arrangement $\mathcal{A}_{\text{shi}(n)}$ are precisely the parking functions of length n, equivalently the superstable configurations of K_{n+1}.*

We see that for any graph G, the maximal superstable configurations are in bijective correspondence with the regions of the graphical arrangement $\mathcal{A}_G$ and that in the special case of the complete graph K_n, the collection of all superstable configurations is in bijective correspondence with the regions of $\mathcal{A}_{\text{shi}(n)}$. We end this section by partially extending the second observation to all graphs.

Similar to the parking functions themselves, the Shi arrangement can be generalized to any graph.

Definition 3.6.7. Given a graph G on n vertices, the *G–Shi arrangement* $\mathcal{A}_{\text{shi}(G)}$ is the hyperplane arrangement in $\mathbb{R}^n$ consisting of the hyperplanes

$$\{x_i = x_j, \quad x_i = x_j + 1\}$$

for all edges $\{i, j\} \in E$, $i < j$.

Mimicking the Pak–Stanley construction, one can label the regions of the G–Shi arrangement. The region labels that are generated are superstable configurations of $G * q$. The correspondence, however, is not bijective. It is surjective. Regions are not assigned unique labels but every G-parking function does appear among the labels. Hopkins and Perkinson [HP16], proving a conjecture of Duval, Klivans and Martin [DKM11], actually prove a more general result. They prove that the Pak–Stanley labels of any *bigraphical arrangement* of a graph yields (surjectively) the superstable configurations of $G * q$.

We illustrate with an example of the G–Shi arrangement in Example 3.6.8.

Example 3.6.8. Figure 3.8 shows the arrangement $\mathcal{A}_{\text{shi}(G)}$ for the graph G equal to the two-edge path. Hence $G * q$ is equal to K_4 minus an edge. There are nine regions in the arrangement, but only eight superstable configurations of $G * q$. A duplicate label can be seen in the two regions just below and to the right and just below and to the left of the center region.

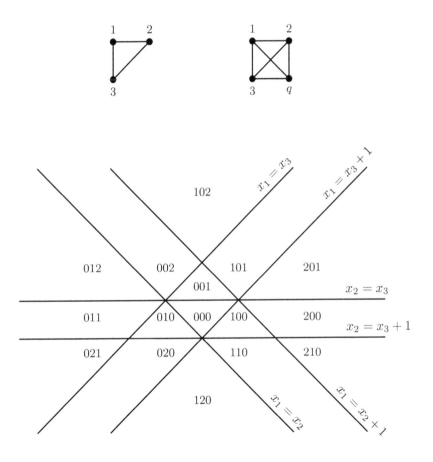

FIGURE 3.7: Shi arrangement with the Pak–Stanley labeling.

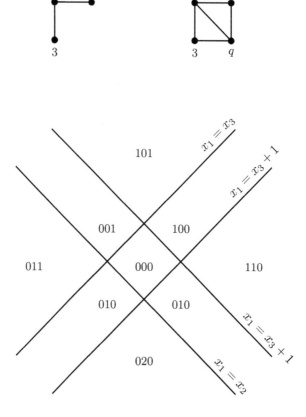

FIGURE 3.8: *G*-version of the Pak–Stanley labeling.

3.7 Dominoes

The theory of domino tilings has a long history in combinatorics and statistical physics. The enumeration of tilings in particular has strong connections to the theory of spanning trees. Below we state a theorem due to Florescu, Morar, Perkinson, Salter and Xu [FMP+14] which uses this connection to show that tilings of the grid are equinumerous with a special kind of critical configuration.

In classical tiling theory, one considers a two-dimensional domain partitioned into unit squares such that two squares meet along an entire edge, at a single point, or not at all. A domino tile is a 2×1 block consisting of two adjacent unit squares. A *domino tiling* is an exact covering of the domain by domino tiles. Equivalently, a domain can be seen as a subgraph of the two-dimensional grid graph, and a domino tiling is a perfect matching or *dimer cover* of the subgraph; see Figure 3.9.

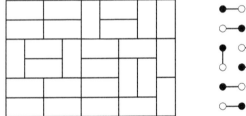 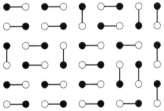

FIGURE 3.9: A two-dimensional domino tiling and corresponding dimer cover. The grid graph is bipartite and we have colored one bipartition of the vertices white and the other black.

Kasteylen showed that domino tilings can be enumerated via a determinant, akin to the Matrix-Tree Theorem. The Matrix-Tree Theorem enumerates spanning trees of a graph via the determinant of the graph Laplacian. The matrix needed here is the Kasteylen matrix defined as follows.

First, let R be the $2m \times 2n$ board. This is one of the simplest domains that one might consider. The parity condition ensures that some tiling exists. Let D be the graph dual to R. The graph D is bipartite and we consider the vertices partitioned into white and black color classes, as in Figure 3.9. An orientation of D is called *admissible* if in each simple cycle of D (local 4-cycle of the grid), the number of edges agreeing with a clockwise orientation of the plane is odd. Kasteleyn proved that such an admissible orientation always exists.

Definition 3.7.1. Let D be the dual graph of a $2m \times 2n$ domain oriented with an admissible orientation. The *Kasteleyn matrix* of D is the black vertices by white vertices adjacency matrix given by

$$K_{ij} = \begin{cases} 1 & \text{if } \{i,j\} \text{ is an oriented edge from black to white} \\ -1 & \text{if } \{i,j\} \text{ is an oriented edge from white to black} \\ 0 & \text{otherwise.} \end{cases}$$

Theorem 3.7.2 ([Kas61]). *Let R be a $2m \times 2n$ domain and D the dual graph of R. The number of perfect matchings of D and hence the number of domino tilings of R is equal to*

$$|\det(K)|.$$

The theory of Kasteleyn matrices is in fact much broader than the theorem above. Admissible orientations and Kasteleyn matrices can be defined for any domain in the plane so that the number of tilings is enumerated by the determinant. Note that without the signs, the number of matchings is given by the permanent of the black to white incidence matrix. Using this theory, one can determine the exact number of tilings of a rectangular board. Although we will not need it, we include the remarkable formula below.

Proposition 3.7.3 ([Kas61],[TF61]). *The number of domino tilings of the $2n \times 2m$ rectangle is:*

$$\left(\prod_{k=1}^{2m} \prod_{j=1}^{2n} \left(4\cos^2 \frac{k\pi}{2m+1} + 4\cos^2 \frac{j\pi}{2n+1} \right) \right)^{\frac{1}{4}}.$$

Next we will see a connection between this enumeration and certain critical configurations of the grid graph.

Let $ST_{m \times n}$ be the $m \times n$ grid graph with a sink added along the boundary. Specifically, the sink is connected to the boundary so that every non-sink vertex has degree exactly four.

Definition 3.7.4. A *symmetric* critical configuration of $ST_{2m \times 2n}$ is a critical configuration that has both horizontal and vertical symmetry.

Theorem 3.7.5 ([FMP$^+$14]). *The number of symmetric critical configurations of $ST_{2m \times 2n}$ is equal to the number of domino tilings of the $2m \times 2n$ board.*

For those readers familiar with the theory of domino tilings, the proof of Theorem 3.7.5 proceeds via the generalized Temperley bijection. The Temperley bijection equates the number of spanning trees of a given grid graph with the number of domino tilings of a grid graph of a different size. For the result of Theorem 3.7.5, the symmetric critical configurations of a given grid are similarly put in bijection with the critical configurations of a grid graph of a different size.

Example 3.7.6. The 4×4 grid has a total of

$$((3 + \sqrt{5})^4 * 3^8 * (3 - \sqrt{5})^4)^{\frac{1}{4}} = 36$$

domino tilings. The 4×4 grid has a total of $100,352$ critical configurations (equiv. spanning trees). Of these, 36 are symmetric; see Figure 3.10, in which the following color coding is used for the number of chips at each site.

Color	Number of chips
■	0
■	1
□	2
■	3

The work of Florescu et al. is broader than Theorem 3.7.5. They consider general group actions on graphs and show that the collection of critical configurations symmetric under a given action forms a subgroup of the sandpile group; see Chapter 4. Furthermore, they use a reduced graph Laplacian corresponding to this subgroup to count symmetric critical configurations in terms of a determinant; see [FMP+14].

3.8 Avalanche polynomials

The early work of Bak, Tang, and Wiesenfeld introduced the concept of self-organized criticality. As we have seen, chip configurations settle to unique critical configurations only through the local firing rule. The *criticality* of the system is that while critical configurations are stable, they are only just barely so. The name is meant to invoke the idea of a critical point of a phase transition. The addition of a few (or just

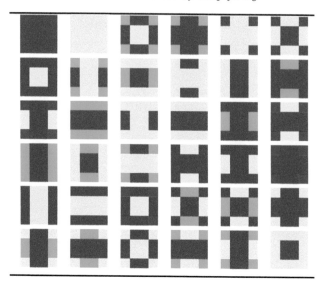

FIGURE 3.10: The 36 symmetric recurrent configurations of the 4×4 grid with a sink along the boundary [FMP$^+$14, Figure 9].

one) chips may cause a critical configuration to become unstable and for firings to occur. Importantly, the addition of a small amount of chips may cause a large amount of firings. This is referred to as *avalanching*.

Bak, Tang and Wiesenfeld observed that in the grid graph avalanche distributions (the number of topplings in an avalanche) obey a power law with exponent approximately -1 (see [BTW88] for a discussion of the connection between criticality and such power laws).

Dhar's subsequent work expanded the chip-firing model to arbitrary graphs. In this expanded domain, do the lengths of topplings still obey a power law? The answer seems to be no in general, but the distributions are nice and take on familiar combinatorial forms. Here we explore the combinatorial power laws generated by different networks.

Suppose **s** is a stable configuration for a graph G on n non-sink vertices and $\mathbf{z} \in \mathbb{Z}_{\geq 0}^n$ represents an arbitrary chip configuration on G. An *avalanche* is any sequence of legal firing moves from $(\mathbf{s} + \mathbf{z})$ to the stabilization stab$(\mathbf{s} + \mathbf{z})$. Recall that the length of any such sequence is the same and hence the length of an avalanche is well defined.

Consider adding a single chip to an already critical configuration.

Definition 3.8.1. Let **c** be a critical configuration on a graph G. A *principal avalanche* is an avalanche formed by stabilizing $(\mathbf{c} + e_i)$ for some i.

Cori, Dartois and Rossin [CDR04] defined the avalanche polynomial to record the lengths of all principal avalanches of a graph.

Definition 3.8.2. The *avalanche polynomial* $Av_G(t)$ of G is the generating function

$$Av_G(t) = \sum_k a_k t^k,$$

where a_k is the number of principal avalanches of length k.

Example 3.8.3. Consider the complete graph on three vertices K_3 with any vertex chosen as the sink. K_3 has three critical configurations. The chart below shows the three critical configurations and the lengths of the two principal avalanches possible from each. For example, the stabilization of $(1,1) + e_1 = (2,1)$ requires 2 chip firings.

critical	e_1	e_2
$(1,1)$	2	2
$(0,1)$	0	1
$(1,0)$	1	0

Therefore the avalanche polynomial is

$$Av_{K_3}(t) = 2t^2 + 2t + 2.$$

The specific form of avalanche polynomials has been computed for only a few families of graphs. We start with complete graphs as in the example above.

For any graph G, we may study the critical configurations of G by instead considering the superstable configurations (via the Duality theorem 2.6.19). For $G = K_n$, the superstable configurations are particularly simple; they are precisely the parking functions of length $(n-1)$. This association allows one to determine avalanche lengths by considering the entries of parking functions. For example, the number of principal avalanches of length 0 on K_n is equal to the total number of non-zero entries in all parking functions of length $(n-1)$. Exploiting the combinatorics of parking functions leads to the following.

Theorem 3.8.4 ([CDR04]). *For the complete graph K_{n+1}, the coefficient a_k of the avalanche polynomial $Av_{K_{n+1}}(t)$ is equal to*

- $n(n-1)(n+1)^{n-2}$ *for $k = 0$*

- $\binom{n}{k} k^{k-1}(n-k+1)^{n-k-1}$ *for* $k > 0$.

Next, let W_n be the wheel graph on $n+1$ vertices consisting of an n-cycle and a center vertex connected to each site on the cycle. Assume that the center vertex is the sink.

Theorem 3.8.5 ([CDR04], [ACF$^+$16]). *Let W_n be the wheel graph as above with $n \geq 3$, then*

$$Av_{W_n}(t) = n^2 t^n + \sum_{m=1}^{n-1} m F_{2(n-m)} t^{m-1} + 2n(F_{2n-1} - 1),$$

where F_n is the nth Fibonacci number.

3.8.1   Avalanche polynomials of trees

Trees present an interesting case; a tree has precisely one critical configuration. And, since $\mathbf{c}_{\max}$ is always critical, it must be the $\mathbf{c}_{\max}$ configuration. The avalanche polynomial thus records the lengths of the principal avalanches of this one critical configuration.

Our first observation is that the avalanche polynomial distinguishes trees from non-trees.

Proposition 3.8.6. *The constant term of the avalanche polynomial vanishes, $Av_G(0) = 0$, if and only if G is a tree.*

Proof. If G is a tree, then there is only one critical configuration, $\mathbf{c}_{\max}$. Adding a chip at site v_i results in a configuration with $\deg(v_i)$ chips at site v_i. If G is not a tree, then G has more than one critical configuration. Let $\mathbf{c}$ be a critical configuration that is not equal to $\mathbf{c}_{\max}$. The configuration $\mathbf{c}_{\max}$ is componentwise maximal, hence there must be some entry i of $\mathbf{c}$ which is strictly less than $\deg(v_i) - 1$. Adding a chip to this site results in a stable configuration, giving at least one principal avalanche of length 0. $\qquad\square$

The avalanche polynomial does not however distinguish between trees. The two trees of Figure 3.11 both have avalanche polynomial equal to

$$Av(t) = t^{10} + t^9 + 6t^8 + 2t^7 + t^4.$$

Austin, Chambers, Funke, Puente and Keough [ACF$^+$16] defined a refinement of the avalanche polynomial which records more information about principal avalanches. For a graph G, assign a variable t_i to each vertex v_i of G. The variable t_i will record how many times v_i fires in an

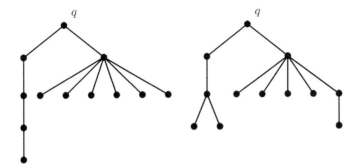

FIGURE 3.11: Two trees with the same avalanche polynomial; see [CDR04].

avalanche. Specifically, for a critical configuration $\mathbf{c}$, define the *avalanche monomial* at v to be the product

$$\mu(\mathbf{c}, v) = \prod_i t_i^{\alpha_i},$$

where α_i is the number of times site i fires in the principal avalanche initiated by adding a single chip to $\mathbf{c}$ at vertex v.

Definition 3.8.7. The *multivariate avalanche polynomial* is defined as

$$Av_G(t_1, \ldots, t_n) = \sum_{\mathbf{c}} \sum_v \mu(\mathbf{c}, v),$$

where the outer sum is over all critical configurations and the inner sum is over all non-sink vertices.

The avalanche polynomial has a useful additivity property.

Proposition 3.8.8. *Let G_1 and G_2 be graphs with a sink. Let $G_1 *_q G_2$ be the graph obtained by identifying the sink vertex of G_1 and the sink vertex of G_2. The avalanche polynomial of the result is*

$$Av_{G_1 *_q G_2}(t) = \tau(G_2) Av_{G_1}(t) + \tau(G_1) Av_{G_2}(t),$$

where $\tau(G_i)$ is the number of spanning trees of G_i.

Using this, one can show that the multivariate avalanche polynomial does distinguish between trees.

Theorem 3.8.9 ([ACF$^+$16]). *If two trees T_1 and T_2 have the same multivariate avalanche polynomial, then T_1 and T_2 are isomorphic.*

Proof. (sketch)

The proof uses two properties of the multivariate avalanche polynomial. The first is the additivity of Proposition 3.8.8. The second is that if a new root is added above the current root, then the polynomial undergoes a simple shift.

With these two properties, the result follows by induction on the height of the tree.   □

3.9   Exercises

Exercise 3.9.1. *Show that for any graph G, the maximal stable configuration $\mathbf{c}_{\max}(G)$ is always critical without using duality (hence showing that the all zeros configuration $\mathbf{0}$ is always a superstable configuration by duality).*

Exercise 3.9.2. *For which graphs is the all zeros configuration a critical configuration?*

Exercise 3.9.3.

1. *Compute the Cori–Le Borgne bijection for all 8 superstable configurations of $K_4 \setminus e$.*

2. *Repeat part 1 for the alternative choice of sink vertex; see G_1 Figure 3.2.*

Exercise 3.9.4. *The duality in Theorem 2.6.19 provides a bijection between critical and superstable configurations. This bijection does not preserve firing-equivalence class. Namely, $\mathbf{c}$ and $\mathbf{c}_{\max} - \mathbf{c}$ are not generally firing-equivalent. Combined with the Cori–Le Borgne bijection this gives an involution on trees. Explore the properties of this involution.*

Exercise 3.9.5. *Let G be a finite graph with edge set E and collection of spanning trees $\mathcal{T}$. Prove that if T_1 and $T_2 \in \mathcal{T}$ and $e \in T_1 \setminus T_2$, then there exists an edge $f \in T_2 \setminus T_1$ such that $(T_1 \setminus e) \cup f \in \mathcal{T}$. Thus proving that the collection of spanning trees of a graph forms the collection of bases of a matroid with ground set E.*

Exercise 3.9.6.

1. Let ∂ be the oriented incidence matrix for a connected graph G. Prove that a collection of columns of ∂ forms a column basis of ∂ if and only if the corresponding collection of edges of G forms a spanning tree of G.

2. Let A be an $k \times n$ matrix and B a $n \times k$ matrix. Prove the Cauchy–Binet formula for computing determinants:

$$\det(AB) = \sum_S \det(A_S)\det(B_S),$$

where the sum runs over all subsets of $\{1, \ldots, n\}$ of size k and A_S (B_S) denotes the matrix formed by restricting to columns (rows) of A (B) indexed by S.

3. Using Parts 1 and 2 give a proof of the Matrix-Tree Theorem, Theorem 3.1.2.

Exercise 3.9.7. *Prove the equivalence of the two definitions of the Tutte polynomial.*

Exercise 3.9.8. *Finish the computation of the Tutte polynomial of $K_4 \backslash e$ by determining the activities of the 8 spanning trees.*

Exercise 3.9.9. *Suppose f is a deletion–contraction invariant of a graph G. Show that f is an evaluation of the Tutte polynomial of G.*

Exercise 3.9.10. *Prove that the Cori–Le Borgne bijection is in fact an activity preserving bijection. In particular, fill in the details of the proof of the first two parts of Theorem 3.4.1 and give a proof for the third claim.*

Exercise 3.9.11. *Suppose there are n cars wanting to park in n parking spots arranged linearly down a street and labeled $0, 1, \ldots, n-1$. Suppose car c_i would like to park in space a_i. The cars come down the street in order: $c_1, c_2, \ldots, c_n$. When car c_i comes down the street, if spot a_i is available, then c_i parks in spot a_i. If the spot a_i is already taken, then car c_i parks in the next available parking spot. If no further spot is available, then car c_i fails to park.*

Define a parking function to be a sequence of preferences $(a_1, a_2, \ldots, a_n)$ that allows all cars to park.

Prove that the parking functions just defined are the parking functions of the complete graph K_n

Exercise 3.9.12. *Prove Theorem 3.6.6 that the Pak–Stanley labeling of the Shi arrangement precisely gives the parking functions of length n.*

Exercise 3.9.13. *Simulate avalanching on the grid graph. Record the lengths of avalanches in order to witness the power law distributions of Bak, Tang and Wiesenfeld.*

Exercise 3.9.14. *Compute the multivariate avalanche polynomials of the two graphs of Figure 3.11.*

Exercise 3.9.15. *(Open)*
  Prove directly that the labels of the G–Shi labeling are componentwise downward closed; see [DKM11].

Chapter 4

Sandpile Groups

Associated with the chip-firing process is an important algebraic invariant – a group structure. The group has gone by various names in various contexts, e.g. the critical group, the Jacobian, the Picard group and the sandpile group. Lorenzini summarizes the approaches in [Lor08]:

- Coming from a physics perspective, Dhar considered grains of sand. The sand avalanches and settles into a system of sandpiles, hence the name the *sandpile* group.

- Biggs studied the chip-firing process as a system of economies and defined the sandpile group under the name the *critical* group.

- Bacher, de la Harpe, and Nagnibeda regarded a graph as a discrete analogue of a Riemann surface. In studying various lattices formed by a graph, they investigated the sandpile group as both the *Jacobian* and *Picard* group of a graph.

- Lorenzini approached chip-firing from the viewpoint of arithmetic geometry using the name the *group of components*.

- The sandpile group can be determined via the Smith normal form of the graph Laplacian. The related group which is determined by the Smith normal form of the adjacency matrix of the graph is known as the Smith group.

We use the terminology *sandpile group* by default. When we want to specifically emphasize a certain perspective, as in Chapter 8 on divisors, we will use the established name of that context.

We start with two different descriptions of the sandpile group, one in terms of chip-firing dynamics and one in terms of the graph Laplacian.

4.1 Toppling dynamics

Suppose G is a finite undirected graph with sink q. Assume that all critical configurations have been normalized so that the value of the sink vertex is equal to minus the sum of the number of chips on the non-sink vertices. Also, in writing out critical configurations explicitly, we will suppress the value of the sink.

Recall that the stabilization operator on a chip configuration $\mathbf{c}$, stab($\mathbf{c}$), yields the unique stable configuration reachable from $\mathbf{c}$ after a sequence of legal chip-firing moves.

Proposition 4.1.1. *Let $\mathbf{c}$ and $\mathbf{d}$ be two critical configurations of G then* stab($\mathbf{c} + \mathbf{d}$) *is also a critical configuration of G.*

Proof. The stabilization of two configurations is stable by definition. For criticality, consider adding the configuration $\mathbf{d}$ one chip at a time; i.e. as a sequence of avalanche operators. Recall that critical configurations are precisely the recurrent configurations of the sandpile Markov chain. Starting at the recurrent configuration $\mathbf{c}$ and performing a sequence of topplings must yield another recurrent configuration.

□

Definition 4.1.2. For two critical configurations $\mathbf{c}$ and $\mathbf{d}$, define the *sandpile sum* $\mathbf{c} \oplus \mathbf{d}$ as follows:

$$\mathbf{c} \oplus \mathbf{d} = \text{stab}(\mathbf{c} + \mathbf{d}).$$

Definition 4.1.3 (Sandpile Group 1). Let G be a finite undirected graph with sink q. The *sandpile group* $\mathcal{S}(G)$ is the finite abelian group on the set of critical configurations of G with addition operator given by the sandpile sum.

Example 4.1.4. Let $G = C_4$ be the four cycle as shown below.

The critical configurations, and thus the elements of the sandpile group of C_4 are:

criticals
$(1,1,1)$
$(0,1,1)$
$(1,0,1)$
$(1,1,0)$

Consider adding two critical configurations as elements of $\mathcal{S}(C_4)$. For example,

$$(0,1,1) \oplus (1,1,0) = \text{stab}(1,2,1).$$

The stabilization of $(1,2,1)$ is shown in Figure 2.7 yielding,

$$(0,1,1) \oplus (1,1,0) = (1,1,0).$$

The remaining sandpile sums are given below:

$$
\begin{array}{rcl}
(1,1,1) \oplus (1,0,1) & = & (1,1,0) \\
(1,1,1) \oplus (0,1,1) & = & (1,1,1) \\
(1,0,1) \oplus (1,0,1) & = & (1,1,1) \\
(0,1,1) \oplus (1,0,1) & = & (1,0,1) \\
(0,1,1) \oplus (0,1,1) & = & (0,1,1) \\
(1,1,1) \oplus (1,1,1) & = & (0,1,1) \\
(1,1,0) \oplus (1,1,1) & = & (1,0,1) \\
(1,0,1) \oplus (1,1,0) & = & (0,1,1) \\
(1,1,0) \oplus (1,1,0) & = & (1,1,1)
\end{array}
$$

From here, one can see explicitly that the sandpile group $\mathcal{S}(C_4)$ is the cyclic group $\mathbb{Z}/4\mathbb{Z}$.

We have defined the sandpile group for a graph with a sink. We will see shortly that the group structure does not depend on the choice of sink. This is in contrast to the critical configurations themselves. The critical configurations may indeed change when the sink is varied (see Example 3.1.4) but the group they form is not changed. Therefore, we can consider the sandpile group $\mathcal{S}(G)$ as an algebraic invariant associated to the graph G.

4.2   Group of chip-firing equivalence

The second formulation of the sandpile group is not as closely tied to the dynamics of chip-firing but has the advantage of being algebraically more straightforward.

Let G be a finite undirected graph with Laplacian Δ. Recall that Δ may be formulated in terms of the oriented incidence matrix of G. We now denote the oriented incidence matrix as ∂_1. In Chapter 2 we used ∂ without a subscript. The subscript reflects the interpretation of ∂_1 as the boundary operator from 1-dimensional to 0-dimensional cells when G is thought of as a 1-dimensional cell complex.[1] The Laplacian of G is then

$$\Delta(G) = \partial_1 \partial_1^T.$$

Let G be a graph on n vertices with sink vertex q. For any chip configuration $\mathbf{c}$ on G, define the map

$$\partial_0 : \mathbb{Z}^n \to \mathbb{Z}, \ \ \partial_0(\mathbf{c}) = \mathbf{c} \cdot \mathbf{1} = c_1 + \cdots + c_{n-1} + c_q.$$

Our earlier convention of normalizing chip configurations so that the value of the sink $\mathbf{c}_q$ is the negation of the sum of all values of $\mathbf{c}$ off the sink ensures that chip configurations are in the kernel of ∂_0.

By definition, the column sums of any graph Laplacian Δ are all zero; i.e. $\Delta \mathbf{1} = \mathbf{0}$. For a connected graph on n vertices, the rank of Δ is equal to $n - 1$. Together this gives a chain complex:

$$\mathbb{Z}^n \xrightarrow{\Delta} \mathbb{Z}^n \xrightarrow{\partial_0} \mathbb{Z} \to 0.$$

The $\operatorname{im}(\Delta) \subseteq \ker(\partial_0)$ because $\Delta \mathbf{1} = \mathbf{0}$. The rank of Δ implies that $\ker(\partial_0)/\operatorname{im}(\Delta)$ is finite.

Definition 4.2.1 (Sandpile Group 2). For a finite undirected graph G the *sandpile group* $\mathcal{S}(G)$ of G is the quotient group

$$\mathcal{S}(G) = \ker(\partial_0)/\operatorname{im}(\Delta).$$

Equivalently, for a graph G with a sink q the sandpile group equals the integer cokernel of the reduced Laplacian:

$$\mathcal{S}(G) \cong \mathbb{Z}^{n-1}/\operatorname{im}(\Delta_q) \cong \operatorname{coker}(\Delta_q).$$

Definition 4.2.1 formulates the sandpile group as a quotient group. Seen as a quotient group, Theorem 2.6.6, which states that there is a unique critical configuration per firing equivalence class, shows that:

The critical configurations (equiv. superstable configurations) of a

---

[1] See Chapter 7 for an introduction to simplicial and cellular complexes.

graph G with sink q can be interpreted as a choice of system of representatives of $\mathcal{S}(G)$.

Either directly from the first definition of the sandpile group or from the second definition along with Theorem 3.1.3 we see that the size of the sandpile group is equal to the number of spanning trees of the graph.

Theorem 4.2.2. *For a finite undirected graph G with sink q, the size of the sandpile group $|\mathcal{S}(G)|$ is equivalently given by:*

1. *The number of critical configurations of G.*

2. *The number of superstable configurations of G.*

3. *The number of spanning trees of G.*

4. *The determinant $\det(\Delta_q(G))$.*

In summary, the sandpile group, $\mathcal{S}(G)$, is an algebraic invariant of G in the form of a finite abelian group whose size equals the number of spanning trees of G.

4.3   Identity

When encountering a group structure, it is natural to consider the identity element of the group. The stabilization operator simply overlays two chip configurations and then topples until stability is reached again. From this perspective, we might envision the all zeros configuration as the identity configuration – adding no chips to an already stable configuration would not alter the configuration.

Upon further consideration, however, we realize that this is almost never possible because the all zeros configuration is almost never a critical configuration; see Section 2.6.1. Notice that the all zeros configuration does not appear in any of our examples of critical configurations. Clearly the identity element must be firing equivalent to the all zeros configuration, but it need not equal the all zeros configuration.

Example 4.3.1. Example 4.1.4 contains the sandpile sum of all pairs of critical elements of $\mathcal{S}(C_4)$. We can see explicitly that the configuration $(0, 1, 1)$ plays the role of the identity element.

The reduced Laplacian for the four cycle is:

$$\Delta_q(C_4) = \begin{array}{c} \\ v_1 \\ v_2 \\ v_3 \end{array} \begin{array}{c} v_1 \quad v_2 \quad v_3 \\ \left(\begin{array}{ccc} 2 & -1 & -1 \\ -1 & 2 & 0 \\ -1 & 0 & 2 \end{array} \right) \end{array}.$$

The configuration $(0, 1, 1)$ is indeed in the image of $\Delta_q(C_4)$:

$$(0, 1, 1)^T = \Delta_q(C_4) \cdot (1, 1, 1)^T,$$

and hence is firing equivalent to the all zeros configuration.

Suppose instead that we consider the sandpile group as a quotient group, with the collection of superstable configurations as our choice of distinguished representatives. Then the all zeros configuration *is* the chosen representative of the identity element. This is because the all zeros configuration is always superstable.

What can be said about the critical configurations that play the role of the identity? Much fascinating behavior has been observed in the identity element of the sandpile group. In particular, identity elements appear to exhibit fractal structure. This behavior will be discussed in detail in Chapter 5. We just give a few intriguing examples here.

Example 4.3.2. Consider the $n \times n$ grid graph with an additional sink vertex q incident to all vertices on the boundary such that each boundary vertex has degree 4. In particular the corner vertices have two edges connected to the sink q.

We have not yet dealt with multiple edges, but firing proceeds as expected: any vertex will need at least 4 chips in order to fire. Firing a corner vertex sends 2 chips to the sink. One can think of the grid as a table top. Chips can fall off the edges to the floor.

The identity elements of the sandpile groups of the 3×3 and 4×4 grid graphs with boundary sink are shown in Figure 4.1.

As we consider larger and larger examples, the representation we have been using becomes somewhat impractical. Working with large grid graphs, chip configurations are often displayed dually with every site represented by a grid square. Below are the identity elements for grids of size 3, 4, and 5.

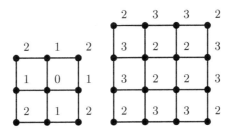

FIGURE 4.1: Identity elements of the 3 × 3 and 4 × 4 grid graphs.

2	1	2
1	0	1
2	1	2

2	3	3	2
3	2	2	3
3	2	2	3
2	3	3	2

2	3	2	3	2
3	2	1	2	3
2	1	0	1	2
3	2	1	2	3
2	3	2	3	2

To recognize some of the intricate behavior of the identity elements, it is helpful to color-code the sites. Suppose we color the sites as indicated in the table below. The color-coded identity elements are shown for larger and larger grids in Figures 4.2, 4.3, 4.4, 4.5 and 4.6.

Color	Number of chips
■	0
■	1
□	2
■	3

FIGURE 4.2: The identity elements for the 3 × 3, 4 × 4, and 5 × 5 grids.

FIGURE 4.3: The identity elements for the 10×10 and 25×25 grids.

FIGURE 4.4: The identity elements for the 50×50 and 100×100 grids.

FIGURE 4.5: The identity element for the 500×500 grid.

FIGURE 4.6: The identity element for the 1000 × 1000 grid.

Already we can see that the identity element seems to be highly symmetric. The larger examples reveal even more fascinating fractal like behavior. Again, we postpone further discussion of these patterns to Chapter 5.

The next proposition gives a method for computing the identity element.

Proposition 4.3.3. *The identity element* $\mathbf{c}_{Id}$ *of the sandpile group is given by*

$$\mathbf{c}_{Id} = \text{stab}(2\,\mathbf{c}_{\max} - \text{stab}(2\,\mathbf{c}_{\max})).$$

Proof. The configuration $\mathbf{c}_{Id}$ is a critical configuration because it is the stabilization of a sufficiently large initial configuration. Furthermore, $\mathbf{c}_{Id}$ is firing-equivalent to the all zeros configuration. Since critical configurations are unique per firing equivalence class, $\mathbf{c}_{Id}$ must be the identity. $\qquad\square$

4.4 Combinatorial invariance

The sandpile group $\mathcal{S}(G)$ of a graph G is an algebraic invariant. If G_1 and G_2 are isomorphic graphs, then the sandpile groups $\mathcal{S}(G_1)$ and $\mathcal{S}(G_2)$ are isomorphic. We can make a more general statement. The next proposition shows that having the same underlying matroid is a sufficient condition to having the same sandpile group.

Proposition 4.4.1 ([Wag00]). *Let G_1 and G_2 be two finite undirected graphs, if their graphical matroids are isomorphic,*

$$M(G_1) \cong M(G_2),$$

then their sandpile groups are isomorphic

$$\mathcal{S}(G_1) \cong \mathcal{S}(G_2).$$

Proof. (sketch) Whitney showed that two graphs have the same graphic matroid if and only if they are related by a sequence of graphical operations known as splittings, mergings, and twistings. Wagner proves that the sandpile group is unchanged under these operations. $\qquad\square$

The converse is false. There exist pairs of graphs with isomorphic sandpile groups but different graphical matroids. Wagner in fact shows something stronger. Two graphs can have the same sandpile group and different Tutte polynomials. An example is given in Example 4.4.2.

Wagner then conjectured that there are graphs with the same Tutte polynomial but non-isomorphic sandpile groups,

Wagner and Merino and de Mier give the following examples.

Example 4.4.2. [Wag00], [dMM01], [GM02].

- Let $C_3 \wedge C_4$ be the graph formed by identifying a vertex of the three cycle with a vertex of the four cycle. The sandpile group is

$$\mathcal{S}(C_3 \wedge C_4) \cong \mathbb{Z}/3\mathbb{Z} \oplus \mathbb{Z}/4\mathbb{Z} \cong \mathbb{Z}/12\mathbb{Z}.$$

The twelve cycle has the same sandpile group:

$$\mathcal{S}(C_{12}) \cong \mathbb{Z}/12\mathbb{Z}.$$

They do not however have the same Tutte polynomial:

$$T_{M(C_3 \wedge C_4)} = (y + x + x^2)(y + x + x^2 + x^3)$$

and

$$T_{M(C_{12})} = y + x + x^2 + \cdots + x^{11}.$$

- Let G_1 and G_2 be the graphs shown below.

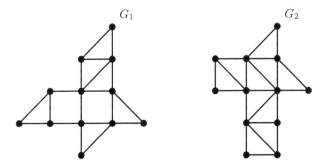

The graphs G_1 and G_2 are not isomorphic, do not have the same sandpile groups, but do have the same Tutte polynomials.

$$
\begin{aligned}
T_{M(G_1)} = T_{M(G_2)} \\
= x^{11} + 10x^9y + 9x^8y^2 + 10x^7y^3 + 12x^6y^4 \\
+ 14x^5y^5 + 14x^4y^6 + 12x^3y^7 + 8x^2y^8 + 4xy^9 + y^10 + 45x^9 \\
+ 81x^8y + 98x^7y^2 + 110x^6y^3 + 118x^5y^4 + 116x^4y^5 \\
+ 97x^3y^6 + 64x^2y^7 + 30xy^8 + 7y^9 + 120x^8 + 288x^7y \\
+ 408x^6y^2 + 462x^5y^3 + 448x^4y^4 + 366x^3y^5 + 236x^2y^6 \\
+ 106xy^7 + 24y^8 + 210x^7 + 588x^6y + 894x^5y^2 + 982x^4y^3 \\
+ 837x^3y^4 + 542x^2y^5 + 240xy^6 + 54y^7 + 252x^6 \\
+ 756x^5y + 1150x^4y^2 + 1158x^3y^3 + 814x^2y^4 + 374xy^5 \\
+ 86y^6 + 210x^5 + 630x^4y + 894x^3y^2 + 762x^2y^3 + 391xy^4 \\
+ 96y^5 + 120x^4 + 336x^3y + 408x^2y^2 + 258xy^3 + 72y^4 \\
+ 45x^3 + 108x^2y + 98xy^2 + 34y^3 + 10x^2 + 18xy \\
+ 9y^2 + x + y.
\end{aligned}
$$

Lorenzini [Lor12] asked similar questions in connection to his two-variable zeta-function. Clancy, Leake, and Payne [CLP15] show, via explicit examples, that none of the following invariants are determined by the other two: the sandpile group, the Tutte polynomial, and the two-variable zeta-function. We defer this result to Chapter 8 when we have developed the necessary material for the zeta-function.

4.5   Sandpile groups and invariant factors

The sandpile group of a graph is a finite abelian group. The fundamental theorem of finite abelian groups states that there exists a decomposition of $\mathcal{S}(G)$ of the following form:

$$
\mathcal{S}(G) = (\mathbb{Z}/n_1\mathbb{Z}) \oplus (\mathbb{Z}/n_2\mathbb{Z}) \oplus \cdots \oplus (\mathbb{Z}/n_r\mathbb{Z})
$$

with the n_i satisfying $n_i|n_{i+1}$ for all i. The n_i are called the *invariant factors* of the group.

As the sandpile group is constructed from the graph Laplacian, it is natural that the invariant factors also appear in a canonical form of the Laplacian.

Definition 4.5.1. For a non-singular integer matrix M, the *Smith normal form* of M is a diagonal matrix A such that

$$A = \mathrm{diag}(d_1, d_1 d_2, \ldots, d_1 d_2 \cdots d_n) = PMQ$$

for invertible matrices $P, Q \in GL(n, \mathbb{R})$ and integers $d_i \in \mathbb{Z}$.

Any non-singular integer matrix M can be put into Smith normal form using the following matrix operations:

Smith normal form

1. Add an integer multiple of one row to another row.

2. Add an integer multiple of one column to another column.

3. Multiply a row or column by ± 1.

Proposition 4.5.2. *If M is a non-singular $n \times n$ integer matrix and the Smith normal form of M is $\mathrm{diag}(e_1, e_2, \ldots, e_n)$ then*

$$\mathrm{coker}_{\mathbb{Z}}(M) \cong (\mathbb{Z}/e_1\mathbb{Z}) \oplus (\mathbb{Z}/e_2\mathbb{Z}) \oplus \cdots \oplus (\mathbb{Z}/e_n\mathbb{Z}).$$

Thus, for a finite abelian group G presented as the cokernel of a non-singular integer matrix M, the invariant factors of G are given by the diagonal entries of the Smith normal form of M. This can be a useful way to compute the invariant factors.

Example 4.5.3. The matrix Δ_q below is the reduced graph Laplacian for $K_4 \backslash e$ with sink at a vertex of degree three. Following the matrix operations above yields the Smith normal form of Δ_q. The invariant factors are 1, 1, and 8 and $\mathcal{S}(K_4 \setminus e) \cong \mathbb{Z}/8\mathbb{Z}$.

$$\Delta_q = \begin{pmatrix} 2 & -1 & 0 \\ -1 & 3 & -1 \\ 0 & -1 & 2 \end{pmatrix} \rightarrow \begin{pmatrix} 2 & -1 & 0 \\ -1 & 2 & -1 \\ 0 & 1 & 2 \end{pmatrix} \rightarrow \begin{pmatrix} 2 & 1 & 0 \\ -1 & 1 & -1 \\ 0 & 1 & 2 \end{pmatrix}$$

$$\rightarrow \begin{pmatrix} 2 & 3 & 0 \\ -1 & 0 & -1 \\ 0 & 1 & 2 \end{pmatrix} \rightarrow \begin{pmatrix} -1 & 0 & -1 \\ 2 & 3 & 0 \\ 0 & 1 & 2 \end{pmatrix} \rightarrow \begin{pmatrix} -1 & 0 & 0 \\ 2 & 3 & -2 \\ 0 & 1 & 2 \end{pmatrix}$$

$$\rightarrow \begin{pmatrix} -1 & 0 & 0 \\ 2 & 3 & -8 \\ 0 & 1 & 0 \end{pmatrix} \rightarrow \begin{pmatrix} -1 & 0 & 0 \\ 2 & 0 & -8 \\ 0 & 1 & 0 \end{pmatrix} \rightarrow \begin{pmatrix} -1 & 0 & 0 \\ 0 & 0 & -8 \\ 0 & 1 & 0 \end{pmatrix}$$

$$\rightarrow \begin{pmatrix} 1 & 0 & 0 \\ 0 & 1 & 0 \\ 0 & 0 & 8 \end{pmatrix}.$$

Proposition 4.5.4 (Theorem of Elementary Divisors). *For a non-singular integer matrix M, let g_i = the gcd of the determinants of all the $i \times i$ minors of M. Then the invariant factors of the integer cokernel of M are equal to:*

$$n_1 = g_1, \ n_2 = \frac{g_2}{g_1}, \ \dots, \ n_r = \frac{g_r}{g_{r-1}}.$$

Let us also point out an enumerative connection to this presentation of the sandpile group. Since the order of the sandpile group of a graph G is equal to the number of spanning trees of G, $\tau(G)$, the invariant factors give a factorization of the tree number:

$$\tau(G) = n_1 n_2 \cdots n_r.$$

The Matrix-Tree Theorem (see Chapter 3 Section 3.1) gives the tree number $\tau(G)$ not only as a determinant but also as a product of eigenvalues. Thus we have two factorizations:

$$|\mathcal{S}(G)| = n_1 n_2 \cdots n_r = \frac{\lambda_1 \lambda_2 \cdots \lambda_{n-1}}{n} = \tau(G).$$

Note that while the invariant factors, the n_i, are always integers, the eigenvalues, the λ_i, can be non-integers.

Example 4.5.5. For the graph $G = K_4 \backslash e$ in the example above, the eigenvalues of the (non-reduced) Laplacian of G happen to be integers. They are $\{4, 4, 2, 0\}$.

$$|\mathcal{S}(G)| = 1 \cdot 1 \cdot 8 = \frac{4 \cdot 4 \cdot 2}{4} = \tau(G) = 8.$$

4.5.1 Explicit forms of the sandpile group

Much work has been done to understand the explicit form (i.e. determine the invariant factors) of the sandpile group for various classes of graphs. Some examples include line graphs [Lev11b], [PSX11], [BMM+12], trees [Tou07], [Lev09], product graphs [JNR03], threshold graphs [CR02], cycles [Lor91], [Mer92] and graphical elliptic curves [Mus09]. In general, this has proved to be a very difficult problem. Here we give just a few examples of the work done in understanding the group structure for various families.

Theorem 4.5.6. *The cycle graph C_n has cyclic sandpile group*

$$\mathcal{S}(C_n) \cong \mathbb{Z}/n\mathbb{Z}.$$

Theorem 4.5.7. *The complete graph K_n has sandpile group*

$$\mathcal{S}(K_n) \cong (\mathbb{Z}/n\mathbb{Z})^{n-2}.$$

Theorem 4.5.7 is a sandpile group version of Cayley's Theorem: the number of spanning trees of the complete graph on n vertices, $\tau(K_n)$, is equal to n^{n-2}.

The proof of both theorems follows from considering the Smith normal form of the reduced Laplacian. The highly symmetric structure of the Laplacian in these cases allows one to determine the invariant factors directly.

Next consider the wheel graph W_n. The graph W_n consists of a cycle of length $n-1$ and one additional vertex connected to every vertex of the cycle. The number of spanning trees of the wheel graph $\tau(W_n)$ is well known especially for the appearance of classical combinatorial sequences:

$$\tau(W_n) = l_{2n} - 2,$$

where l_k is the kth Lucas number. The Lucas numbers are recursively defined and can be expressed as a sum of Fibonacci numbers,

$$l_k = f_{k-1} + f_{k+1}.$$

Explicitly,

$$l_k = \left(\frac{1 + \sqrt{5}}{2}\right)^n + \left(\frac{1 - \sqrt{5}}{2}\right)^n.$$

Theorem 4.5.8 ([Big99a], [NW11]). *Let $G = W_n$ be the wheel graph on n vertices and $\tau = \tau(W_n)$. Then*

if n is even

$$\mathcal{S}(G) \cong \mathbb{Z}_{\sqrt{\tau}} \times \mathbb{Z}_{\sqrt{\tau}},$$

and
if n is odd

$$\mathcal{S}(G) \cong \mathbb{Z}_{\sqrt{\tau/(n-4)}} \times \mathbb{Z}_{\sqrt{\tau/(n-4)}}.$$

In general, one might ask which finite abelian groups arise as sandpile groups. The next proposition allows us to see that every finite abelian group appears as the sandpile group of some graph.

Suppose G is a graph which can be obtained from graphs $G_1, G_2, \ldots, G_k$ by gluing all the G_i along a single vertex v. Then v is called an *articulation point* of G and the G_i are the *articulated components* of G with respect to v.

Proposition 4.5.9 ([CR00]). *The sandpile group $\mathcal{S}(G)$ of a graph G with articulation point v and articulated components $G_1, \ldots, G_k$ is the product of the sandpile groups of the articulated components:*

$$\mathcal{S}(G) \cong \mathcal{S}(G_1) \times \mathcal{S}(G_2) \times \cdots \times \mathcal{S}(G_k).$$

Proof. Under the assumptions of the proposition, the Laplacian of G is block diagonal with one block per articulated component.   □

Corollary 4.5.10. *Every finite abelian group is the sandpile group of some planar graph.*

Proof. Consider a finite abelian group presented in terms of its invariant factors. For each cyclic summand, take the cycle graph of appropriate size (as in Theorem 4.5.6). Identify the cycles along a single vertex and apply Proposition 4.5.9.[2]   □

Therefore every finite abelian group appears as the sandpile group of at least one graph. In the next section, we consider the frequency with which various groups appear as the sandpile group across all graphs.

4.5.2 Sandpile groups of random graphs

Interesting questions arise not just from looking at specific classes of graphs but by considering the form of sandpile groups in general. Both Wagner and Lorenzini made observations about the overall structure of sandpile groups, especially their tendency to frequently be cyclic.

One way to consider the general form of sandpile groups is through a probabilistic lens:

What does the sandpile group of a random graph look like?

What is the probability that the sandpile group of a random graph is cyclic?

---

[2]For the group $\mathbb{Z}/2\mathbb{Z}$ we need the non-simple graph consisting of two vertices and two edges.

Lorenzini [Lor08, Section 4] proposed three different measures to consider in asking such questions. Here, we will only look at the Erdös–Renyei random graph model.

Definition 4.5.11. An *Erdös–Renyei random graph* $H \in G(n,p)$ is a graph on n vertices generated by including each possible edge independently with probability p.

Wagner [Wag00] conjectured that for a large range of edge probabilities p, as n increases, the probability that the sandpile group is cyclic converges to 1. The solution turns out to be more delicate.

Understanding the structure of sandpile groups benefits not only from the perspective of random graphs, but also the theory of random groups and random matrices. We will state a result conjectured by Clancy, Leake, and Payne and proved by Wood but without proof as the techniques are sufficiently out of scope for this text.

As in Wagner's conjecture, we will be interested in the limiting distribution of sandpile groups of Erdös–Renyei random graphs as n grows. Unfortunately, there is little information to be gained from asking about the precise form of the sandpile group as seen in the next proposition due to Wood [Woo17].

Proposition 4.5.12. *For a fixed finite abelian group* $\mathcal{G}$ *and* $H \in G(n,p)$,

$$\lim_{n \to \infty} \mathrm{prob}(\mathcal{S}(H) \cong \mathcal{G}) = 0.$$

However, one can quantify the distribution of the Sylow p-subgroups of $\mathcal{S}(H)$.

The sandpile Sylow p-subgroups follow a distribution related to the *Cohen–Lenstra heuristic*. The Cohen–Lenstra heuristic (or principle) arose in number theory in the context of distributions of class groups of number fields. But, the idea has turned out to be more universal. The Cohen–Lenstra heuristic concerns the distribution of finite abelian groups as witnessed in a variety of contexts.

The principle states that within some naturally arising collection of finite abelian groups, a fixed group $\mathcal{G}$ appears with frequency proportional to one over the number of automorphisms of the group:

$$\frac{1}{|\mathrm{Aut}(\mathcal{G})|}.$$

And indeed this seems to be true for a variety of settings.

The Sylow p-subgroups of random sandpile groups do not follow this exact heuristic. Instead, as was initially observed in [CLP15], sandpile groups follow a variation of the Cohen–Lenstra heuristic that includes the notion of a *duality pairing*.

Definition 4.5.13. A *duality pairing* on a finite abelian group $\mathcal{G}$ is a symmetric, bilinear, perfect mapping $\langle \cdot, \cdot \rangle \to \mathbb{Q}/\mathbb{Z}$ such that the map $\mathcal{G} \to \mathrm{Hom}(\mathcal{G}, \mathbb{Q}/\mathbb{Z})$ given by $g \to \langle g, \cdot \rangle$ is an isomorphism.

Bosch and Lorenzini [BL02] showed that there is a canonical duality pairing for the sandpile group of a graph G with sink q and Laplacian Δ given by

$$\langle x, y \rangle_G = y^T \Delta_q^{-1} x.$$

For a graph G with pairing $\langle \cdot, \cdot \rangle_G$, a finite abelian group $\mathcal{G}$ appears as a Sylow subgroup of $\mathcal{S}(G)$ with frequency proportional to

$$\frac{1}{|\mathcal{G}| |\mathrm{Aut}(\mathcal{G}, \langle \cdot, \cdot \rangle_G)|},$$

where $\mathrm{Aut}(\mathcal{G}, \langle \cdot, \cdot \rangle_G)$ denotes the subgroup of automorphisms of $\mathcal{G}$ that preserves the duality pairing.

Theorem 4.5.14 ([Woo17]). *Let p be a prime and $\mathcal{G}$ a finite abelian p-group. Then for a random graph $H \in G(n, q)$, with $\mathcal{S}(H)_p$ the Sylow p-subgroup of the sandpile group of H,*

$$\lim_{n \to \infty} \mathrm{prob}(\mathcal{S}(H)_p \cong \mathcal{G}) =$$

$$\frac{|\text{symmetric, bilinear, perfect maps } \phi : \mathcal{G} \times \mathcal{G} \to \mathbb{C}^*|}{|\mathcal{G}||\mathrm{Aut}(\mathcal{G})|} \prod_{k \geq 0} (1 - p^{-2k-1}).$$

Using arithmetic techniques, this theorem allows one to gain understanding of the structure of sandpile groups. For example one has

Corollary 4.5.15 ([Woo17]). *For $H \in G(n, q)$, the probability that $\mathcal{S}(H)$ is cyclic is bounded as follows:*

$$\lim_{n \to \infty} \mathrm{prob}(\mathcal{S}(H) \text{ is cyclic}) \leq \zeta(3)^1 \zeta(5)^1 \zeta(7)^1 \zeta(9)^1 \zeta(11)^1 \cdots \equiv .793$$

where ζ is the Riemann zeta function.

The distribution of Theorem 4.5.14 explains the prevalence of cyclic sandpile groups at an even finer level. Theorem 4.5.14 implies that the group $\mathcal{G} = (\mathbb{Z}/p^r\mathbb{Z})$ appears with frequency p^{-r} while the group $\mathcal{G} = (\mathbb{Z}/p\mathbb{Z})^r$ appears with frequency $p^{-r(r+1)/2}$, so for example, $\mathbb{Z}/49\mathbb{Z}$ is the Sylow 7-subgroup of a sandpile group approximately 7 times more often than $(\mathbb{Z}/7\mathbb{Z})^2$ is the Sylow 7-subgroup.

4.6 Discriminant groups

In this section we give another formulation of the sandpile group as the discriminant group of the cut and flow lattices of a graph as first shown in [BdlHN97]; see also [GR01].

Recall the construction of the graph Laplacian as the product $\partial\partial^T$, where ∂ is the vertex by edge matrix of oriented incidences. These oriented incidence maps also reflect the cuts and flows of a graph.

Definition 4.6.1. The *cut and flow spaces* and *cut and flow lattices* of an oriented graph G are

$$\mathrm{Cut}(G) = \mathrm{im}_{\mathbb{R}}\,\partial^T, \qquad\qquad \mathrm{Flow}(G) = \ker_{\mathbb{R}}\partial,$$
$$\mathcal{C}(G) = \mathrm{im}_{\mathbb{Z}}\,\partial^T, \qquad\qquad \mathcal{F}(G) = \ker_{\mathbb{Z}}\partial.$$

How do these spaces reflect the standard notions of cuts and flows?

For a finite undirected graph $G = (V, E)$, a collection of edges C is a *cut* if there exists a partition $V = \{U, W\}$, such that each edge in C has one endpoint in U and one endpoint in W.

For an oriented graph and an ordered partition $V = (U, W)$, the *signed characteristic vector* of the cut C induced by $\{U, W\}$, $\chi(C)$, records the orientation of each edge of C with respect to the ordering of the partition:

$$\chi(C)_e = \begin{cases} 1 & \text{if } e \text{ is oriented from } U \text{ to } W, \\ -1 & \text{if } e \text{ is oriented from } W \text{ to } V, \\ 0 & \text{otherwise.} \end{cases}$$

The columns of ∂^T can be interpreted as signed characteristic vectors of cuts: For each vertex v, partition V as $(v, V \setminus v)$. The induced cut consists of all edges incident to v. Moreover, the cut space $\mathrm{Cut}(G)$ contains the signed characteristic vectors of all cuts of G.

The flow space is the orthogonal complement of the cut space. Let $C^1(G; \mathbb{R})$ denote the vector space of real-valued functions on the edges of G, then

$$C^1(G; \mathbb{R}) = \mathrm{Cut}(G) \oplus \mathrm{Flow}(G).$$

The simple cycles of G form the minimal elements of the flow space.

Let G be an oriented graph. Let C be a cycle of G and F the collection of edges of the cycle C. Consider traversing the edges of F in one of the two cyclic orderings induced by C. The *signed characteristic vector*, $\chi(F)$, records the orientation of each edge along the traversal with respect to the orientation of the edges in G.

$$\chi(F)_e = \begin{cases} 1 & \text{if } e \text{ is traversed in the same orientation it has in } G, \\ -1 & \text{if } e \text{ is traversed in the opposite orientation it has in } G, \\ 0 & \text{otherwise.} \end{cases}$$

The flow space $\text{Flow}(G)$ contains the signed characteristic vectors of all flows of G.

Example 4.6.2. Consider $K_4 \backslash e$ labeled as shown. Suppose all edges have been directed from smaller to larger label.

$$\partial_{K_4 \backslash e} = \begin{array}{c} \\ 1 \\ 2 \\ 3 \\ 4 \end{array} \begin{pmatrix} \begin{array}{ccccc} 12 & 13 & 14 & 24 & 34 \\ 1 & 1 & 1 & 0 & 0 \\ -1 & 0 & 0 & 1 & 0 \\ 0 & -1 & 0 & 0 & 1 \\ 0 & 0 & -1 & -1 & -1 \end{array} \end{pmatrix}.$$

The two edges incident to vertex 3 give a minimal cut C of the graph. The vertex set V is partitioned as $V = (3, 124)$, and

$$C = \{13, 34\}.$$

$$\partial^T_{K_4 \backslash e} = \begin{array}{c} \\ 12 \\ 13 \\ 14 \\ 24 \\ 34 \end{array} \begin{pmatrix} \begin{array}{cccc} 1 & 2 & 3 & 4 \\ 1 & -1 & 0 & 0 \\ 1 & 0 & -1 & 0 \\ 1 & 0 & 0 & -1 \\ 0 & 1 & 0 & -1 \\ 0 & 0 & 1 & -1 \end{array} \end{pmatrix}.$$

The characteristic vector of the cut C is

$$\chi(C) = \begin{array}{c} \begin{array}{ccccc} 12 & 13 & 14 & 24 & 34 \end{array} \\ \left(\begin{array}{ccccc} 0 & -1 & 0 & 0 & 1 \end{array} \right). \end{array}$$

By definition, a flow is any element of the kernel of ∂. The flow

$$F = (1, 0, 1, -1, 0),$$

corresponds to a simple cycle in the graph as shown below.

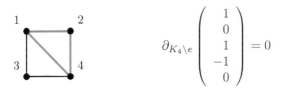

$$\partial_{K_4 \setminus e} \begin{pmatrix} 1 \\ 0 \\ 1 \\ -1 \\ 0 \end{pmatrix} = 0$$

Next we consider the cut and flow lattices. The group

$$\mathbb{Z}^{|E|} / (\mathcal{C} \oplus \mathcal{F})$$

is called the *cutflow* group of G.

In order to arrive at the sandpile group, we need the notion of a lattice dual.

Definition 4.6.3. For a lattice L, the *dual lattice* of L is

$$L^\sharp = \{ v \in L \mid \langle v, w \rangle \in \mathbb{Z} \ \forall w \in L \}.$$

Note that $(L^\sharp)^\sharp = L$. A lattice is called *integral* if it is contained in its dual, $L \subseteq L^\sharp$.

Definition 4.6.4. The *discriminant group* (or *determinantal group*) of an integral lattice L is the quotient

$$\mathrm{Disc}(L) = L^\sharp / L.$$

Bacher, de la Harpe, and Nagnibeda studied the lattices $\mathcal{C}$ and $\mathcal{F}$ of integral cuts and flows for a graph in [BdlHN97]. They interpreted the discriminant groups $\mathcal{C}^\sharp / \mathcal{C}$ and $\mathcal{F}^\sharp / \mathcal{F}$ respectively as the Picard group of divisors and as the Jacobian group of holomorphic forms. We will explore this perspective in Chapter 8 on divisors on graphs. Here we follow the presentation of [GR01].

The following lattice theoretic facts allow one to show that the discriminant groups, cutflow group, and sandpile group are all isomorphic.

Proposition 4.6.5. *Let M be an $n \times r$ integer matrix. Suppose the columns of M form an integral basis for the lattice L.*

1. *Then*
$$|L^\sharp/L| = \det(M^T M).$$

2. *The columns of $M(M^T M)^{-1}$ form the corresponding dual basis for $L^\sharp$.*

3. *The matrix $P = M(M^T M)^{-1} M^T$ represents orthogonal projection from $\mathbb{R}^n$ onto the column space of M.*

4. *If the greatest common divisor of the $r \times r$ minors of M is 1, then $L^\sharp$ is generated by the columns of P.*

Let $M = \partial^T$. Then the matrix P from Proposition 4.6.5 is a projection to the cut space and is an isomorphism from $\mathbb{Z}^{|E|}/(\mathcal{C} \oplus \mathcal{F})$ to $\mathcal{C}^\sharp/\mathcal{C}$. The matrix $I - P$ which projects to the flow space gives an isomorphism from $\mathbb{Z}^{|E|}/(\mathcal{C} \oplus \mathcal{F})$ to $\mathcal{F}^\sharp/\mathcal{F}$.

Theorem 4.6.6. *For a finite graph G, the discriminant groups of the cut and flow lattices of a graph are isomorphic to each other, to the sandpile group of G, and to the cutflow group of G:*

$$\mathbb{Z}^{|E|}/(\mathcal{C} \oplus \mathcal{F}) \cong \mathcal{C}^\sharp/\mathcal{C} \cong \mathcal{F}^\sharp/\mathcal{F} \cong \mathcal{S}(G).$$

One corollary to Theorem 4.6.6 is a particularly elegant explanation of the isomorphism between sandpile groups of planar graphs and their duals.

Corollary 4.6.7. *Let G be a planar graph and G^* its planar dual. Then*

$$\mathcal{S}(G) \cong \mathcal{S}(G^*).$$

Proof. For a planar graph G, the flow space of G is isomorphic to the cut space of G^* making their cutflow groups isomorphic. $\square$

Corollary 4.6.7 is established more directly in [CR00].

Example 4.6.8. Let $G = K_4 \backslash e$ and G^* the planar dual to G. It is easily checked that both G and G^* have sandpile group $\mathbb{Z}/8\mathbb{Z}$.

4.7   Sandpile torsors

Our starting point in this section is once again the fact that the size of the sandpile group is equal to the number of spanning trees of a graph. One would like to find an explicit relationship between the group and the set of spanning trees. The bijections of Chapter 3 connect critical and superstable configurations to trees and even preserve activity.

Is it possible to construct a bijection from the set of spanning trees to the sandpile group in a *natural* way? The naturality condition should mean (at least) that the map only depends on the isomorphism class of G.

Wagner formalized the question as follows. For two graphs G and H, let $\mathcal{T}(G)$ and $\mathcal{T}(H)$ denote the respective collections of spanning trees. If $f : G \to H$ is an isomorphism then f induces a bijection $f_\mathcal{T}$ between $\mathcal{T}(G)$ and $\mathcal{T}(H)$. The graph isomorphism f also induces a group isomorphism $f_\mathcal{S}$ between the sandpile groups $\mathcal{S}(G)$ and $\mathcal{S}(H)$:

$$f_\mathcal{T} : \mathcal{T}(G) \to \mathcal{T}(H)$$
$$f_\mathcal{S} : \mathcal{S}(G) \to \mathcal{S}(H).$$

The goal then is to construct maps ψ_G and ψ_H from the collection of trees to the sandpile group such that the following diagram commutes.

$$
\begin{array}{ccc}
\mathcal{T}(G) & \xrightarrow{\psi_G} & \mathcal{S}(G) \\
{\scriptstyle f_\mathcal{T}} \downarrow & & \downarrow {\scriptstyle f_\mathcal{S}} \\
\mathcal{T}(H) & \xrightarrow{\psi_H} & \mathcal{S}(H)
\end{array}
$$

Unfortunately there are graphs for which there are no such isomorphisms [Wag00, Theorem 8.1].

Informally, we can recognize that the set of spanning trees of a graph does not naturally exhibit a group structure. In particular, the set of spanning trees does not have a distinguished element (namely an element to play the role of the identity).

Instead, we might ask if there is a nice *action* of the sandpile group on the set of spanning trees.

Definition 4.7.1. Let $\mathcal{G}$ be a finite group and X a set with equal cardinality. The set X is a *torsor* for $\mathcal{G}$ if there is a free and transitive action of $\mathcal{G}$ on X. Namely, for any two elements $x, y \in X$, there exists a unique element $g \in \mathcal{G}$ such that $gx = y$.

In our case the group in question is the sandpile group $\mathcal{S}(G)$ of a graph G and the set of equal cardinality is the set of all spanning trees $\mathcal{T}(G)$.

For a set X and a group $\mathcal{G}$, there may be many torsor structures, i.e. free transitive actions of $\mathcal{G}$ on X.

We will describe three distinct torsor actions for the sandpile group $\mathcal{S}(G)$ and the set of spanning trees $\mathcal{T}(G)$. An interesting aspect of this story is that each uses a different combinatorial structure to provide the action – rotor-routing, tree transversals, and cycle–cocycle reversals from matroid theory – yet the torsor structures seem to be closely related.

Each of the three constructions requires choices on the graph such as the choice of a sink.

The first two actions we describe require the input of a ribbon graph.

Definition 4.7.2. A *ribbon graph* (or *combinatorial embedding*) is a graph G along with a choice of cyclic ordering of the edges incident to each vertex.

4.7.1 Rotor-routing

Rotor-routing is a process for the diffusion of chips on a graph that is different from chip-firing. In rotor-routing, as a chip arrives at a vertex, it is immediately fired to a neighboring vertex. The sharing of chips is kept as equal as possible by rotating through the neighbors of vertices. We will not explore the full theory of the rotor-routing process, only define what we need for a torsor structure; see [HLM$^+$08] for more on rotor-routing.

Suppose we are given a ribbon graph. The cyclic orderings at each vertex are the *rotors*. At each vertex, a choice of outgoing edge gives the state of the rotor. In rotor-routing, chips move from vertex to vertex according to the state of the rotor – if, at any time, a chip is at a vertex v, it is *routed* to a neighboring vertex according to the chosen outgoing edge. Furthermore, each time a chip is moved from v to an adjacent vertex, the rotor state (outgoing edge) is first updated to the next edge in the cyclic ordering given by the ribbon graph structure.

To define a torsor action using rotor-routing, start with a ribbon graph G with a sink q. Note that $\mathcal{S}(G)$ is generated by the collection of configurations which have value -1 at q, value 1 at a single non-sink vertex and 0 elsewhere. Therefore, given a spanning tree T of G, we define an action of the sandpile group by specifying the tree gT for configurations g of the above form.

Given a spanning tree T of a graph G with a sink, there is a unique orientation of the edges of T *towards* the sink. This collection of oriented edges gives a rotor configuration of G at all vertices except at the sink.

Let g be a chip configuration with value 1 at v, -1 at the sink q and 0 elsewhere. First update the rotor at v, i.e. replace the outgoing edge of the spanning tree T at v with the next edge in the cyclic order at v. Second, move the single chip at v along the updated rotor edge to an adjacent vertex v'. Repeat this process until the single chip is moved to the sink q. When the single chip reaches the sink, the rotor configuration is again a tree oriented towards the sink. The tree, denoted $T_{[v]}$, is the image of the action on T by the configuration g; see Example 4.7.5.

Procedure *Rotor-routing torsor(G, T)*
1. Input a ribbon graph G and spanning tree T
2. Fix a sink q of G.
3. Orient the edges of the tree toward the sink.
4. Place a chip on a vertex v and follow the rotor-routing process until the chip is at q.
5. Output the resulting rotor configuration.

Theorem 4.7.3 ([HLM+08]). *Rotor-routing gives a torsor structure for the sandpile group on a ribbon graph with a sink.*

Interestingly, the choice of sink does not matter in the case of planar graphs.

Proposition 4.7.4 ([CCG15]). *The rotor-router torsor is independent of the choice of sink if and only if the graph is planar.*

Example 4.7.5. Consider $K_4 \backslash e$ with sink q as in Figure 4.7. Assume that the ribbon structure at each vertex is clockwise oriented with respect to the embedding in the figure.

Let T_1 be the set of bold edges in the first graph of the top row. Suppose one chip has been placed at vertex v_3. Following the rotor-routing procedure moves the single chip to the sink in three moves. The induced tree is labeled T_2:

$$(0, 0, 1, -1) \cdot T_1 = T_2.$$

In the second row, the sink has changed and the input tree is now T_2.

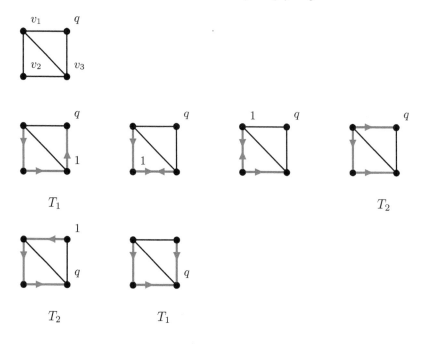

FIGURE 4.7: An example of the rotor-routing torsor action. The cyclic ordering at each vertex is clockwise.

In one step the chip is moved into the sink and the induced tree is T_1. Therefore

$$(0, 0, -1, 1) \cdot T_2 = T_1$$
$$(0, 0, -1, 1)^{-1} \cdot T_1 = T_2$$
$$(0, 0, 1, -1) \cdot T_1 = T_2.$$

The change of sink does not affect the torsor structure because of the planarity of the ribbon graph $K_4 \backslash e$.

4.7.2 Bernardi process

The Bernardi process is another torsor structure for a ribbon graph with sink. Baker and Wang [BW17] introduce this sandpile torsor based on tree bijections constructed by Bernardi [Ber08].

The sandpile group $\mathcal{S}(G)$ defined as $\ker(\partial_0)/\operatorname{im}(\Delta)$ consists of chip configurations whose total sum over all vertices is equal to zero. Let $\mathcal{S}^g(G)$ denote the collection of equivalence classes of firing-equivalent chip configurations whose total sum over all vertices is equal to g. (So the sandpile group is $\mathcal{S}^0(G)$.)

The Bernardi process proceeds in two steps. First, the spanning trees of a graph G can be put in bijective correspondence with $\mathcal{S}^g(G)$. Second, the set $\mathcal{S}^g(G)$ can itself be seen as a torsor for $\mathcal{S}(G)$. Combining the two constructions gives a free transitive action on the set of spanning trees.

For the first step, we need the notion of a break configuration. For a finite undirected graph G, the *genus* of G is

$$g = m - n + 1,$$

where m is the number of edges and n is the number of vertices of G. Let T be a spanning tree of a graph G with genus g, then T has precisely $n-1$ edges and hence g edges not in T. A *break configuration* of G is any chip configuration which consists of g chips total and which is formed by choosing a spanning tree T of G and placing, for each edge $e \notin T$, one chip at one of the endpoints of e; see Figure 4.8.

Proposition 4.7.6 ([MZ08], [ABKS14]). *For a graph G with genus g, every chip configuration on G with a total of g chips is firing-equivalent to a unique break configuration.*

In analogy to both the collection of critical and superstable configurations, we interpret Proposition 4.7.6 as saying that break configurations form a system of representatives for $\mathcal{S}^g(G)$.

Both Backman [Bac17] and Yuen [Yue18] have given algorithms to check whether or not a configuration is a break configuration.

Example 4.7.7. Consider the graph $K_4 \setminus e$. The genus of $K_4 \setminus e$ is $g = 5 - 4 + 1 = 2$. Figure 4.8 shows two different choices of spanning tree, represented as blue edges. For each tree, two examples of break configurations are shown. The configurations consist of $g = 2$ chips. Notice that the second and fourth break configurations are the same even though the underlying spanning trees are not.

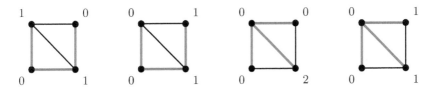

FIGURE 4.8: Examples of break configurations.

Next we describe the bijection between spanning trees and break configurations. Let G be a ribbon graph, q a choice of sink of G and e a choice of edge incident to q. Let T be a spanning tree of G.

With this input data, traverse (each side of) the edges of T as follows. Start at the vertex q. If e is in T, walk along the edge e. If e is not in T, walk along the first edge incident to q appearing after e in the cyclic ordering at q (given by the ribbon graph structure). Continue traversing the edges in T by following the cyclic ordering of the ribbon graph structure at each vertex. Form a chip configuration on G by placing a chip at a vertex each time the walk cuts through an edge not in T; see Figure 4.9.

Let $\beta_{(q,e)}(T)$ be the chip configuration resulting from this process.

Proposition 4.7.8 ([Ber08], [BW17]). *The map* $\Phi : \mathcal{T}(G) \to \mathcal{B}(G)$ $\Phi(T) = \beta_{(q,e)}(T)$ *is a bijection between spanning trees and break configurations.*

Example 4.7.9. Consider again $K_4 \setminus e$ as shown in Figure 4.9. The cyclic ordering at each vertex is clockwise. The sink is q and the choice of edge e is dashed. In the second row, a spanning tree T has been chosen, dark edges are in T. Starting at q, we walk down the edge e and then follow the clockwise order at each vertex to arrive back at q. Both gray edges (edges not in T) are first crossed at the top left vertex. Thus the top left vertex gets two chips.

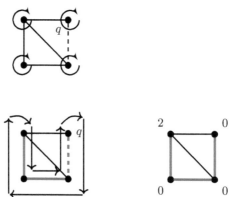

FIGURE 4.9: An example of the Bernardi bijection.

The sandpile group $\mathcal{S}^0(G)$ has a simple transitive action on $\mathcal{S}^g(G)$. For configurations $\mathbf{c} \in \mathcal{S}^0(G)$ and $\mathbf{d} \in \mathcal{S}^g(G)$, the sum, $\mathbf{c} + \mathbf{d}$ remains in $\mathcal{S}^g(G)$. By Proposition 4.7.6, every element of $\mathcal{S}^g(G)$ is firing equivalent to a unique break configuration. Hence we have a simple transitive action of $\mathcal{S}^0(G)$ on break configurations by adding a zero-sum configuration to a break configuration and then returning the break configuration that is equivalent to the sum.

We are now ready to describe the Bernardi torsor action. Given

- $h \in \mathcal{S}^0(G)$ an element of the sandpile group and

- $T \in \mathcal{T}(G)$ a spanning tree,

let $B \in \mathcal{S}^g(G)$ be the break configuration such that $\Phi(T) = B$. Let

$$B' \in \mathcal{S}^g(G) \text{ be firing equivalent to } B + h.$$

Let T' be the spanning tree of G such that $\Phi(T') = B'$. Then the Bernardi process is the action

$$h \cdot T = T'.$$

Theorem 4.7.10 ([Ber08] [BW17]). *The Bernardi process is a torsor structure for the sandpile group on a ribbon graph with sink.*

Interestingly, Baker and Wang further show that for planar graphs, the Bernardi and rotor-routing torsors coincide. Note that this implies that the Bernardi torsor is also independent of the choice of sink in the case of planar graphs.

Baker and Wang further conjecture that this is the only case in which they coincide. Namely, they conjecture that for any non-planar graph, there is always some choice of sink such that the Bernardi and rotor-routing actions are not the same.

Break configurations will be considered again in Chapter 8 where they are referred to as *break divisors*. In particular, they will index a polyhedral subdivision of the degree g Picard torus.

4.7.3 Cycle–cocycle reversal

Another torsor structure is introduced in [BBY17]. In this setting, the action of the sandpile group is generalized from spanning trees of graphs to bases of all regular matroids. Regular matroids include graphical matroids, and we will focus on the graphical case.

For this torsor structure we need the set of *cycle–cocycle reversal equivalence classes* as first introduced by Gioan [Gio07]. For a graph G, the cycle–cocycle reversal relation is a relation on the set of all orientations of G. A *cycle reversal* operation reverses the orientation of all edges in a directed cycle (positive circuit in the oriented matroid). A *cocycle reversal* operation reverses the orientation of all edges in a directed cut. A directed cut (directed cocircuit in the oriented matroid) is a cut in which all edges have the same orientation from one side of the induced vertex partition to the other. Two orientations of G are related if one can be obtained from the other via a sequence of cycle and cocycle reversals. This defines an equivalence relation on all orientations of a graph.

Example 4.7.11. Consider the four cycle, C_4. There are a total of $2^4 = 16$ orientations of the four cycle C_4. Figure 4.10 shows the possible orientations of C_4 partitioned by equivalence class. Two orientations have an arrow between them if they are related by a single cycle or single cocycle reversal.

Proposition 4.7.12 ([Gio07]). *For any graph G, the number of equivalence classes of the cycle–cocycle reversal equivalence relation is equal to the number of spanning trees of G.*

Gioan proves this via a deletion/contraction method showing that the number of equivalence classes is equal to the evaluation of the Tutte polynomial at $(1, 1)$. In fact he shows more: for any regular matroid M, the number of equivalence classes of the circuit–cocircuit reversal equivalence relation is equal to the number of bases of M.

Moreover, the cycle–cocycle reversal classes form a torsor structure for the sandpile group [Gio07], [Bac17]. For this action, recall from the previous section that the sandpile group $\mathcal{S}(G)$ is isomorphic to the cutflow group $\mathbb{Z}^{|E|}/(\mathcal{C} \oplus \mathcal{F})$ and elements of the group can thus be represented by classes of linear combinations of oriented edges.

For a graph G, let e be an oriented edge of G and $\mathcal{O}$ an orientation of the collection of edges of G. The action of the sandpile group on cycle–cocycle orientation classes is given as follows: if the orientation of e is the same as the orientation of e in $\mathcal{O}$, then reverse the edge e in $\mathcal{O}$ to produce $\mathcal{O}'$. If the orientation of e is not the same as the orientation of e in $\mathcal{O}$, then perform cycle–cocycle reversals on $\mathcal{O}$ until the orientations are the same. Then, reverse the edge e to produce $\mathcal{O}'$.

In order to define a torsor structure on spanning trees, Backman, Baker, and Yuen [BBY17] construct a family of bijections between spanning trees and cycle–cocycle classes (and more generally between bases of a regular matroid and circuit–cocircuit classes). As in the Bernardi case, given a bijection between spanning trees and reorientation classes, one can get a torsor structure on trees by the sandpile group by using the action of the cycle–cocycle classes. Alternatively, by fixing a class

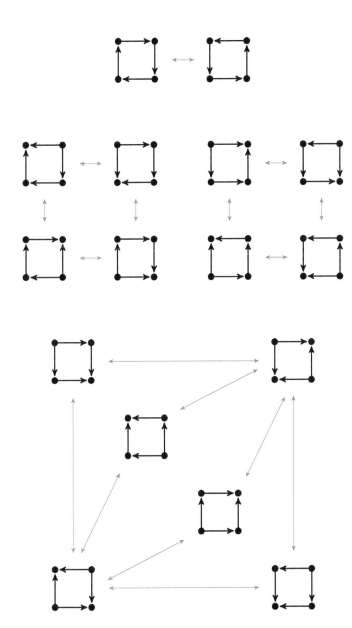

FIGURE 4.10: The four equivalence classes of the cycle–cocycle reversal relation on C_4.

in the cycle–cocycle reversal system to correspond to the identity in the sandpile group, one can get a family of bijections between spanning trees and the sandpile group.

For a graph G, the bijections given in [BBY17] begin by fixing an *acyclic cycle signature* σ and an *acyclic cut signature* σ^* for G. These are certain orientations of the simple cycles and simple cuts of G.

We describe one way to obtain such acyclic orientations for a graph G. First, fix both a linear order of all edges and an orientation of each edge of G. For each simple cycle F in G, let $\sigma(F)$ be the orientation of F compatible with the orientation of the smallest edge in F. Similarly, for each simple cut C in G, let $\sigma^*(C)$ be the orientation compatible with the orientation of the smallest edge in C.

Construct a map Φ from spanning trees of G to orientations on the set of edges of G. Given a spanning tree $T \subset G$, define an orientation of G,

$$\mathcal{O}(G) = \Phi(T),$$

as follows:

$$\mathcal{O}(e) = \begin{cases} \text{the same as in } \sigma(C(T,e)) & \text{if } e \notin T \\ \text{the same as in } \sigma^*(C^*(T,e)) & \text{if } e \in T, \end{cases}$$

where $C(T,e)$ is the fundamental cycle of e with respect to T and $C^*(T,e)$ is the fundamental cut of e with respect to T; see Chapter 3 Section 3.2.3.

Theorem 4.7.13 ([BBY17]). *The map Φ is a bijection from spanning trees to cycle–cocycle reorientation classes.*

Example 4.7.14. Let G be the 4-cycle with vertex and edge labels as shown in Figure 4.11. There is only one simple cycle of G:

$$[e_1 + e_3 - e_4 - e_2].$$

There are six simple cuts of G:

$$[e_1 - e_3], [e_3 + e_4], [e_2 - e_4],$$

$$[e_1 + e_2], [e_2 + e_3], [e_1 + e_4].$$

Figure 4.11 shows the four spanning trees of G along with the orientation given by the map Φ. Comparing to Figure 4.10, we see one orientation from each cycle–cocycle class in the image of Φ.

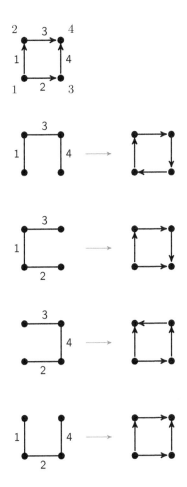

FIGURE 4.11: An example of the bijection Φ between spanning trees and cycle–cocycle classes.

For example, consider the tree T consisting of edges $\{1, 3, 4\}$. The edge 2 is not in T, therefore we orient 2 the same as its orientation in the fundamental cycle containing 2. The fundamental cycle containing 2 is the entire cycle, which has been oriented consistently with its smallest element, the edge 1. Therefore we see the edge 2 directed from right to left.

The four orientations given in the image of Φ are all distinct under the cycle–cocycle reorientation relation; see Figure 4.10.

We are now ready to describe the cycle–cocycle torsor action. Given

- $h \in \mathcal{S}(G)$ an element of the sandpile group and

- $T \in \mathcal{T}(G)$ a spanning tree,

let CC be the cycle–cocycle orientation such that $\Phi(T) = CC$. Compute

$$CC' = h \cdot CC$$

Let T' be the spanning tree of G such that $\Phi(T') = CC'$. Then the cycle–cocycle process is the action

$$h \cdot T = T'.$$

Theorem 4.7.15 ([BBY17]). *The cycle–cocycle process is a torsor structure for the sandpile group for a graph with choice of acyclic cycle signature.*

Proposition 4.7.16 ([BBY17]). *For planar graphs, the cycle–cocycle reorientation torsor coincides with the Bernardi and rotor-routing torsor and hence is independent of the sink.*

Proposition 4.7.16 leads to the following conjecture.

Conjecture 4.7.17. *For planar graphs, there is only one sandpile torsor structure.*

Contrived examples easily break Conjecture 4.7.17. For this problem, one would first need to define a class of suitably nice torsor structures.

As previously mentioned, the results of Backman, Baker and Yuen are more general and apply to all regular matroids. Furthermore, we have given only one specific example of a bijection between trees and orientation classes. The paper [BBY17] sets up a family of elegant bijections using the geometry of zonotopal subdivisions.

That a sandpile torsor is independent of the sink in the planar case can be interpreted as: the torsor can distinguish when the topological genus of a graph is 0. The results of [McD18] show that this essentially cannot be extended to higher genus.

4.8 Exercises

Exercise 4.8.1. *Show that the sandpile group, as defined in Definition 4.1.3 is in fact a group. Namely, show that the critical configurations along with the sandpile sum $\oplus$ satisfy the axioms of a group.*

Exercise 4.8.2. *Prove the claim in the construction of the sandpile group as a quotient:*

$$\Delta \mathbf{1} = 0 \implies \mathrm{im}(\Delta) \subseteq \ker(\partial_0).$$

Exercise 4.8.3. *Show the equivalence of the sandpile group in Definition 4.2.1 and with respect to the sink:*

$$\mathcal{S}(G) = \ker(\partial_0)/\mathrm{im}(\Delta) \cong \mathbb{Z}^{n-1}/\mathrm{im}(\Delta_q) \cong \mathrm{coker}(\Delta_q).$$

Conclude that the sandpile group is independent of the choice of sink q.

Exercise 4.8.4. *Prove that any integer matrix always has a Smith normal form. More generally prove that any matrix over a PID has a Smith normal form.*

Exercise 4.8.5. *Compute the sandpile group, i.e. the invariant factors, of the 3×3 grid, the 4×4 grid, the $n \times n$ grid.*

Exercise 4.8.6. *Prove that the sandpile group of a tree is trivial.*

Exercise 4.8.7. *Prove that the sandpile group of the n-cycle is cyclic for all $n \geq 3$, $\mathcal{S}(C_n) \cong \mathbb{Z}/n\mathbb{Z}$.*

Exercise 4.8.8. *Prove that the sandpile group of the complete graph K_n is $(\mathbb{Z}/n\mathbb{Z})^{n-2}$.*

Exercise 4.8.9. *Fill in the details of the proof of Proposition 4.3.3: The identity element $\mathbf{c}_{Id}$ of the sandpile group is given by*

$$\mathbf{c}_{Id} = \mathrm{stab}(2\,\mathbf{c}_{\max} - \mathrm{stab}(2\,\mathbf{c}_{\max})).$$

Exercise 4.8.10. *Determine the sandpile groups of the two graphs G_1 and G_2 of Example 4.4.2.*

Exercise 4.8.11. *Prove that the cut and flow lattices of a graph are integral lattices.*

Exercise 4.8.12. *Alternate basis for the flow space. For a graph G, let T be a fixed spanning tree of G. Adding one edge not already in T to T defines a unique cycle, a fundamental circuit; see Section 3.2.3. Prove that the collection of signed characteristic vectors of the collection of fundamental circuits forms a basis for the flow space of G.*

Exercise 4.8.13. *Alternate basis for the cut space. For a graph G, let T be a fixed spanning tree of G. Removing one edge at a time from T defines a unique cut, a fundamental cocircuit; see Section 3.2.3. Prove that the collection of signed characteristic vectors of the collection of fundamental cocircuits forms a basis for the cut space of G.*

Exercise 4.8.14. *Prove the claims preceding Theorem 4.6.6.*

1. *The matrix P is an isomorphism from the cutflow group to the discriminant group of the cut lattice.*

2. *The matrix I − P is an isomorphism from the cutflow group to the discriminant group of the flow lattice.*

Exercise 4.8.15. *Using Exercise 4.8.14, prove Theorem 4.6.6.*

Exercise 4.8.16. *Prove that for a graph G with sink q, the sandpile group is generated by all configurations of the form: −1 chips at q and 1 chip at v at all v ≠ q.*

Exercise 4.8.17. *Let G be the 3 × 3 grid. Fix a sink vertex and a spanning tree of G. Place a single chip on a non-sink vertex and run the router-routing process to produce a new spanning tree as in the rotor-routing torsor.*

Exercise 4.8.18. *Prove Proposition 4.7.6, that for a graph of genus g, every chip configuration with g chips is firing-equivalent to a unique break configuration.*

Exercise 4.8.19. *Prove Proposition 4.7.8, that the Bernardi bijection is in fact a bijection between spanning trees and break configurations.*

Exercise 4.8.20. *Let G be a graph with genus g. Let B ≥ 0 be a configuration with g chips total. Show that B is a break configuration if and only if for every connected subgraph H of G, the number of chips contained in the restriction of B to H is at least the genus of H.*

Chapter 5

Pattern Formation

5.1 Compelling visualizations

Early on in the study of sandpile dynamics, sandpile simulations revealed fascinating configurations. Bak [BTW88], Dhar [Dha06], Creutz [Cre04] and Ostojic [Ost03] seem to have created some of the first simulations of sandpile processes. See Creutz [Cre04] for a fun description of having the office across the hall from Bak and the excitement of visualizations in the time of the new microcomputer.

Here we discuss early observations and experiments and more recent theoretical results concerning pattern formation.

In Chapter 1, our first examples were of small configurations on small grid graphs. In our imagery of sand falling on a planar surface, these were toy examples. When considering a granular flow, one would want as fine a grid as possible and as many grains of sand as possible.

Figures 5.1 and 5.2 show the stabilization of the initial configuration which consists of 10 thousand, 10 million and 20 million chips at the origin and 0 chips elsewhere on the grid $\mathbb{Z}^2$. For these figures, recall our color-coding scheme. A stable configuration on the grid has at most 3 chips at each site. The table below shows the assignment of a distinct color to each height.

Color	Number of chips
■	0
■	1
□	2
■	3

Clearly an amazing pattern is forming as we observe the result of stabilizing larger and larger initial configurations of chips at the origin. (Due to earlier considerations we could also think of forming this

FIGURE 5.1: The stabilization of 10 thousand and 10 million chips at the origin.

FIGURE 5.2: The stabilization of 20 million chips at the origin.

configuration by adding one chip at a time and stabilizing between each addition.) The apparent fractal nature of the stabilization is particularly intriguing.

Fractal like patterns have been observed in other setups as well. Figure 5.15 for example demonstrates a fractal pattern formed via chip-firing from a single initial stack of chips on a planar lattice of degree 8.

Figure 5.21 shows a configuration on a large grid graph. The configuration of Figure 5.21 is not the stabilization of an initial stack of chips at the origin. In this case, the underlying grid graph is finite and has a sink. The sink is not shown but is connected to every boundary site. The configuration displayed is the identity element of the sandpile group of this graph as represented by a critical configuration. Namely, it is the configuration that is both critical and firing equivalent to the all zeros configuration; see Chapter 4 Section 4.3.

Our formal understanding of these patterns and others is quite limited. But the observed patterns have spurred much experimental work and some convergence results. We will only take an abbreviated look at the theory of pattern formation; we refer to [Pao14] for a more comprehensive treatment of the topic.

5.2   Infinite graphs

The first three chapters of the text worked solely with finite graphs. In this chapter, we expand to the case of infinite graphs. We allow the number of vertices of a graph to be infinite but assume that the degree of each vertex is finite.

In the *chip-firing process on an infinite graph*:

- A chip configuration specifies the number of chips at each vertex.

- An initial configuration may consist of an infinite number of chips, but there must be only a finite number of chips at each vertex.

- The firing rule remains the same. A vertex is ready to fire if it has at least as many chips as it has neighbors. A legal fire is one in which no entry of the result is negative.

- The chip-firing process on an infinite graph starts with an initial configuration. At each time step, a vertex that is ready to fire is selected and fired. If, at any stage, a stable configuration is reached, the process stops.

- Local confluence, i.e. the diamond property continues to hold. If at some time, two vertices can each legally fire, then they can legally fire in either order and the resulting configuration is the same (see Theorem 2.2.2).

For an initial configuration **c** on G, define the *support* of **c** to be the collection of all sites in G that fire during the chip-firing process with initial configuration **c**.

Proposition 5.2.1. *Any initial configuration on a connected infinite graph with only finitely many chips will eventually stabilize.*

Proof. Let G be a connected infinite graph and **c** an initial chip configuration on G consisting of finitely many chips.

Suppose vertex v fires during the chip-firing process starting at **c**. Then in all future steps of the process, for each neighbor u of v, either u or v must have at least one chip; simply consider the most recent of the two to fire.

Furthermore, suppose that a subset of vertices W have all fired at some point during the chip-firing process starting at **c**. Then in all future steps of the process, the number of chips on vertices in W is at least the number of edges in the induced subgraph on W. If **c** has finitely many chips, this implies that only finitely many vertices will ever fire in the chip-firing process starting at **c**, i.e. **c** has finite support.

If **c** did not stabilize, some vertex w of the support would have to fire infinitely often. Then every vertex in the same connected component of the support as w would also have to fire infinitely many times. The connectedness of G implies that there exist vertices of the connected component of w with neighbors outside of the support. Each fire of these vertices would send at least one chip outside of the support. Once outside the support, these chips will never fire again. With only finitely many chips to begin with, this process must stop. $\qquad\square$

In contrast, initial configurations with infinitely many chips may not stabilize. A simple example is the $\mathbb{Z}^2$ grid with 4 or more chips at each site. Fey, Levine and Peres [FLP10] give the following example with very few chips at each vertex in the initial configuration. Consider the $\mathbb{Z}^2$ grid. Place 4 chips at the origin, 3 chips at each location along the x-axis, and 2 chips at all other locations. Initially, only the origin can legally fire. This one firing will, however, cause infinitely many firings to occur. A further interesting aspect to this example is that although infinitely many firings will occur, each site will only fire finitely many times.

Most of our focus for this chapter will be a very simple chip-firing scenario. Let G be the graph of the $\mathbb{Z}^d$ lattice for some d. Let $\delta_0(n)$ be the configuration with n chips at the origin and zero chips elsewhere. This configuration is known as the *pulse* at the origin. We will be interested in the stabilization of the pulse, $\text{stab}(\delta_0(n))$.

5.3 The one-dimensional grid

Anderson, Lovasz, Shor, Spencer, Tardos and Winograd [ALS+89] analyzed the one-dimensional case as a combinatorial game (good for "an interesting diversion during long lectures or faculty meetings"). It remains the only fully understood case, although perhaps this heavy hitting collection of authors just need to find themselves sufficiently bored once again.

Let G be the graph of the one-dimensional lattice $\mathbb{Z}$ and $\delta_0(n)$ the discrete pulse at the origin. We observed above that two adjacent sites which both fire during the stabilization of the pulse cannot both have value 0 in $\text{stab}(\delta_0(n))$.

This observation limits the structure of $\text{stab}(\delta_0(n))$. The support must lie between $-2n$ and $2n$. The degree of each vertex is 2, so the number of chips at each vertex in the stabilization is either 0 or 1. Which pattern of 0s and 1s (without two adjacent 0s) is equal to $\text{stab}(\delta_0(n))$? The fractal images in $\mathbb{Z}^2$ might lead us to think we would see something interesting in the locations of the 0s, e.g. perhaps they would follow an arithmetic progression.

A few examples prove otherwise. Consider $\text{stab}(\delta_0(5))$. Figure 5.3 shows the stabilization in three steps. At each step, we have performed multiple legal fires simultaneously including firing a vertex as many times as legally possible.

Theorem 5.3.1 ([ALS+89]). *For an initial pulse of n chips at the origin of the line, $\delta_0(n)$, we find the following behavior.*

- *For $n = 2k + 1$, $\text{stab}(\delta_0(n))$ consists of exactly one chip at each site in the interval $[-k, k]$.*

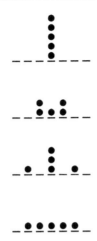

FIGURE 5.3: The stabilization stab($\delta_0(5)$) of 5 chips at the origin on the one-dimensional grid.

- *For $n = 2k$, stab($\delta_0(n)$) consists of exactly one chip at each site in the set $[-k, -1] \cup [1, k]$.*

Proof. First note that by the confluence property, it is enough to show that there exists a sequence of legal fires that reaches the desired final configurations.

Let $n = 2k + 1$. We sketch the sequence followed in [ALS$^+$89]. Consider the symmetric configurations consisting of l chips at the origin where l is odd and one chip in the next m consecutive positions: $[1, m]$ and $[-m, -1]$. Fire the center vertex once. Now perform all possible legal fires at all locations except the origin. Continue to perform all successive fires at all locations except at the origin. When there are no legal firing moves except at the origin, the configuration will have l chips at the origin, zero chips at locations 1 and -1, and one chip in the next m consecutive positions: $[2, m + 1]$ and $[-(m + 1), -2]$; see Figure 5.4. The runs of consecutive ones have been pushed out one step from the origin. Firing the center vertex once yields the configuration similar to our starting configuration but with two less chips at the origin and one chip each at locations $(m + 1)$ and $-(m + 1)$. Repeating this process will eventually produce the configuration with one chip at each site in the range $[-k, k]$.

For the case $n = 2k$, start with the odd case. In the construction above, one chip never needs to move. It can stay at the origin for the entire stabilization process. Imagine removing this chip and otherwise following the same firing sequence as above. The final configuration is the same with one less chip at the origin.

□

Following the firing sequence in the proof of Theorem 5.3.1, the firing sequence that occurs between fires of the origin is a cascade rippling through the sequence of ones until the boundary expands and then a chip cascades back to the origin. This is demonstrated below by indicating the number of chips at each location at each step; see also Figure 5.4.

	1	1	1	1	3	1	1	1	1	
	1	1	1	2	1	2	1	1	1	
	1	1	2	0	3	0	2	1	1	
	1	2	0	2	1	2	0	2	1	
	2	0	2	0	3	0	2	0	2	
1	0	2	0	2	1	2	0	2	0	1
1	1	0	2	0	3	0	2	0	1	1
1	1	1	0	2	1	2	0	1	1	1
1	1	1	1	0	3	0	1	1	1	1
1	1	1	1	1	1	1	1	1	1	1

Recall that local confluence (the diamond lemma) also guarantees that in any stabilization sequence, the length of the sequence and number of times each site fires is the same.

Theorem 5.3.2 ([ALS$^+$89]). *In any stabilization of $\delta_0(2k)$ or $\delta_0(2k+1)$ on the one-dimensional grid,*

1. *The total number of fires is $\frac{k(k+1)(2k+1)}{6}$.*

2. *Site i (and $-i$) fires exactly $\binom{k-i+1}{2}$ times for $0 \leq i \leq (k-1)$.*

Proof. For part 2, let $f(i)$ be the number of times site i fires in the stabilization of $\delta_0(n)$. Then $f(i)$ satisfies the recursion

$$f(i) = \lfloor \frac{f(i-1)}{2} \rfloor + \lfloor \frac{f(i+1)}{2} \rfloor \text{ for } i \neq 0$$
$$f(i) = \lfloor \frac{f(i-1)}{2} \rfloor + \lfloor \frac{f(i+1)}{2} \rfloor + \frac{n}{2} \text{ for } i = 0.$$

The size of the support gives the boundary condition for when $f(i)$ becomes non-zero. The triangle numbers $\binom{k-i+1}{2}$ uniquely satisfy the recursion and boundary condition.

Part 1 is simply the sum of the first k triangle numbers and first $k-1$ triangle numbers.

□

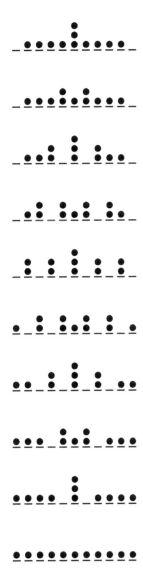

FIGURE 5.4: The cascading behavior of the one-dimensional pulse.

5.4   Labeled chip-firing

Before moving on to two dimensions, we look at a variant of the one-dimensional case introduced by Propp.

Let G be the graph of the one-dimensional lattice. Consider n *labeled* chips at the origin. Suppose the chips are labeled $\{1, \ldots, n\}$.

At any site with at least two chips, choose any two chips and fire them to neighboring sites by sending the chip with the larger label one step to the right and the chip with the smaller label one step to the left. At each step, if any site has at least two chips continue firing until no location has more than one chip.

As with standard unlabeled chip-firing, we are immediately led to ask about the order of firings. Does it matter which two chips are selected to fire first? By simply erasing the labels on the chips, we know that the final chip configuration will be (for n even) a single chip at locations $\{-k, -(k-1), \ldots, -1\}$ and $\{1, 2, \ldots, k\}$ and 0 chips elsewhere.

We consider a few examples.

Example 5.4.1. Figure 5.5 shows an initial configuration of four labeled chips at the origin and two possible paths to stabilization. (Not all paths to stabilization are shown.) In one firing move, the initial configuration can reach either configuration shown in the second row. There does **not** exist a configuration reachable from these two intermediate configurations in a **single** fire. After two more fires, there is a configuration reachable from both.

The example shows that the labeled chip-firing process can violate the local confluence property (see Theorem 2.2.2).

For the two paths shown here, global confluence is not violated. And, it is not hard to show that for this initial configuration, the final configuration shown in Figure 5.5 is the unique reachable stable configuration.

Example 5.4.2. Figure 5.6 shows an initial configuration of five labeled chips at the origin and two possible paths to stabilization. (Not all paths to stabilization are shown.)

This example shows that the labeled chip-firing process can violate not only the local confluence property but also global confluence (again see Theorem 2.2.2).

Examples 5.4.1 and 5.4.2 show that labeled chip-firing does not fall into our earlier regimes. Local confluence is lost. Hopkins, McConville, and Propp show that for *even* values of n, *global* confluence is maintained.

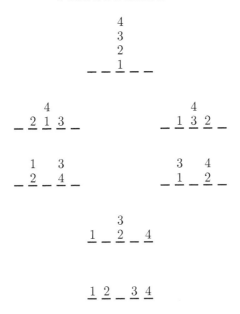

FIGURE 5.5: Local confluence (the diamond property) is violated in this example of labeled chip-firing.

Theorem 5.4.3 ([HMP17]). *For $n = 2k$ labeled chips at the origin, there is a unique final stable configuration in the labeled chip-firing process. Furthermore, in the final configuration, the chips are always sorted, appearing from smallest to largest.*

Theorem 5.4.3 only holds for $n = 2k$ chips. The conclusion does not hold for an odd number of chips. Indeed, we can easily see this failure already at $n = 3$. Suppose chips labeled 1, 2, and 3 are placed at the origin. Fire the chips labeled 2 and 3. The resulting configuration is $\underline{213}$. This configuration is stable and so the process terminates unsorted.

Although the odd case does not sort, there is a remarkable conjecture for the odd case. Suppose one implements labeled chip-firing for $n = 2k + 1$ chips starting at the origin of the line. Consider the following three randomized strategies for determining a firing sequence:

- At each step, choose a legal firing move uniformly at random from all possible legal fires.

- At each time step, choose a vertex that is ready to fire uniformly at random from all vertices ready to fire and then choose a pair of chips from the vertex also uniformly at random.

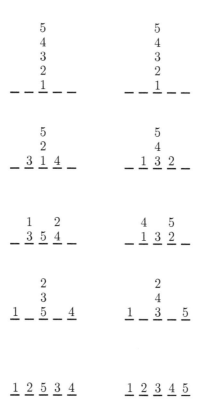

FIGURE 5.6: The final configuration on the left is not sorted. The final configuration on the right is sorted

- Choose a stabilizing firing sequence uniformly at random from all possible sequences.

Conjecture 5.4.4 ([HMP17]). *Under any of the three procedures above, the probability of resulting in a sorted configuration tends to* $\frac{1}{3}$.

An interesting observation about this conjecture comes from considering the last vertex to fire. In any stabilization sequence, the last fire always occurs at a vertex with exactly three chips. There are precisely three sequences to stabilization from a vertex with three chips. Two of these fail to sort, as in our first non-sorting example. Therefore the conjecture is claiming that with probability tending to 1, labeled chip-firing sorts the chips up to the last step.

The proof of Theorem 5.4.3 is surprisingly involved. What makes this problem significantly harder than the unlabeled counterpart is the loss of local confluence. The global confluence must be established without using local confluence.

In order to form the labeled analogy to Theorem 5.3.2, we consider a variation of labeled chip-firing where only two labels are used. Suppose each chip is colored red or blue. When two chips of the same color fire, send one left and one right. When two chips of different colors fire, send the red chip one step to the left and the blue chip one step to the right. We say that this colored chip-firing sorts if, independent of the firing sequence, in the final configuration, all red chips are to the left of all blue chips.

Lemma 5.4.5. *Labeled chip-firing, with n labels, sorts if and only if colored chip-firing sorts for all possible two-colorings of the chips.*

Proof. Assume that colored chip-firing sorts.

Now consider the labeled chip-firing process. Consider two labels $i < j$. Color all chips with label at most i red and the remaining chips blue. A stabilizing sequence of labeled chip-firing can be seen as a stabilizing sequence of colored chip-firing. Then under our assumption, the chip labeled i must end to the left of the chip labeled j.

For the other direction, suppose we have n chips colored blue and red. Label the chips with unique labels such that all red chips get a label smaller than all blue chips.

Any stabilizing sequence of colored chip-firing can be seen as a stabilizing sequence of labeled chip-firing: when a red and blue chip fire, the chip of smaller label moves left and the chip of larger label moves right. When two chips of the same color fire, colored chip-firing makes no distinction, simply chose the chip of smaller label to move left. Since the red chips all have smaller labels than the blue chips, the fact that labeled chip-firing sorts implies the red chips are to the left of the blue chips in the final configuration.   □

Hence an alternative proof for Theorem 5.4.3 would be to prove that two-colored chip-firing with $n = 2k$ chips results in a sorted final configuration. This simplification of the problem unfortunately does not seem to make the solution any simpler.

The total number of fires is not changed with the introduction of labels or colors and is given in Theorem 5.3.2. Let RB be the number of of times a red chip fires with a blue chip. The quantity RB is a color-sensitive refinement of the total number of fires.

Theorem 5.4.6 ([FK17]). *For an initial configuration of $n = 2k$ red and blue chips at the origin, let n_r be the number of red chips, n_b be the number of blue chips and $m = \min(n_r, n_b)$. Then*

$$\text{RB} = \sum_{i=k-m+1}^{k} i.$$

Proof. Let

$$Z = \sum_{i=k-m+1}^{k} i.$$

We in fact show something stronger than the claim of the proposition. For any value of n, RB $\leq Z$. And, for $n = 2k$, RB $= Z$ if and only if colored chip-firing sorts.

Let S_r be the sum over all red chips of the position of each chip. Similarly let S_b be the sum over the blue chips of the position of each chip. In the starting configuration $S_r = S_b = 0$. When two chips of the same color fire, both S_r and S_b remain unchanged. When a red and a blue chip fire together, S_r decreases by one, and S_b increases by one. Therefore S_r is non-increasing, S_b is non-decreasing, $S_r = -\text{RB}$ and $S_b = \text{RB}$.

Suppose $m = n_b$. (The case where $m = n_r$ is analogous.) The maximum final value of S_b is obtained when the blue chips are as far right as possible, in positions $\{k - m + 1, \ldots, k\}$. In this case, $S_b = Z$. Moreover, this is the only final configuration with $S_b = Z$. We conclude RB $\leq Z$ and RB $= Z$ exactly when all red chips are to the left of all blue chips in the final configuration.   □

Not only does this give us a colored invariant, the proof gives another alternative route to the proof of Theorem 5.4.3.

For unlabeled chip-firing, the total number of fires *at each site* is also an invariant to the firing sequence. It is not the case however that the number of red–blue fires at each site is an invariant in colored chip-firing.

The idea of labeled chip-firing has been extended by Galashin, Hopkins, McConville and Postnikov to a concept of *root system chip-firing*. In this work, the firing rule is reinterpreted as follows. For a labeled chip configuration on $\mathbb{Z}$ consisting of chips $\{1, \ldots, n\}$, define a vector x by setting x_i equal to the position of the chip with label i. Now chips i and j ($i < j$) can fire if x is orthogonal to $(e_j - e_i)$. The resulting configuration corresponds to the vector $x + (e_j - e_i)$.

The collection of vectors $\{e_j - e_i\}$ for $i < j$ is the collection of positive roots of the type A root system. With this starting point, one can investigate chip-firing on the weight lattice of any root system under the firing rule

$$\lambda \to \lambda + \alpha$$

where α is a positive root orthogonal to a weight λ.

Again, the first question one asks is whether this chip-firing process is confluent. It is shown that this process is confluent modulo the action of the associated Weyl group. Unfortunately, in type A modding out by the action of the Weyl group corresponds to forgetting the labels of the chips. In the labeled case, which systems are confluent from which weights is intricate; see [GHMP17a] and [GHMP17b].

An introduction to root systems and an alternative construction for chip-firing on weight lattices appears in Section 6.6. This alternative chip-firing process for root systems is an example of a more general setup which is always locally and globally confluent.

Generally, it appears to be a difficult problem to prove global confluence of a process without the diamond lemma as a tool. In [DF91], Diaconis and Fulton investigate confluence in connection to a wide array of ideas including Lagrange inversion and characteristic classes of projective spaces. In a series of works, [BL16a, BL16b, BL16c], Bond and Levine have introduced *abelian networks* as axiomatic systems that precisely satisfy a confluence property. Abelian networks have been studied particularly from the context of finite-state automata [HLW15].

5.5   Two and more dimensional grids

The most attention in chip-firing pattern formation has been on the two-dimensional grid.

We repeat our setup of n chips at the origin and 0 chips elsewhere. Let G be the graph of the two-dimensional grid $\mathbb{Z}^2$. Let $\delta_0(n)$ denote the configuration with n chips at the origin and 0 chips elsewhere. Each site in the stabilization of $\delta_0(n)$ can take on one of four possible values $0, 1, 2$ or 3.

Consider the configuration $\text{stab}(\delta_0(n))$. The one-dimensional case might lead us to think that the chips will simply spread out, as evenly as possible, symmetrically in all dimensions. Chips do not however simply spread out to a constant height in some domain centered at the origin. Experimentally, it quickly becomes clear that there is much more complicated behavior in two dimensions than in the one-dimensional case.

Computations reveal intriguing patterns. Figures 5.1 and 5.2 represent the stabilization of 10-thousand, 10-million, and 20-million chips at the origin. We can see visually that there appears to be a fractal like pattern, spreading to an almost circular region, with hyperbolic-like triangular regions of constant height.

Although one can easily "see" these patterns, it is quite challenging to prove precise statements about the structure of the configurations. First, we set up the formalization of the problem.

Let s_n denote the stabilization $\text{stab}(\delta_0(n))$. Thus we have:

$$s_n = \delta_0(n) + \Delta v_n, \tag{5.1}$$

where

- s_n is a configuration on $\mathbb{Z}^2$ which will be non-zero only at finitely many sites.

- v_n is a function from $\mathbb{Z}^2$ to $\mathbb{Z}$ known as the *odometer* function. The value of the ith entry of v_n represents the total number of times that site i fires during the stabilization process.

- Δ is the discrete Laplace operator on the grid. We must be a little careful here. Since we are working on an infinite grid and not a finite graph, Δ can no longer be seen as a finite matrix. Instead, Δ is the point operator:

$$\Delta u(x) = \sum_{y \sim x} u(y) - u(x),$$

where x is a site in $\mathbb{Z}^2$ and the sum is over all neighbors of x.

For dimensions two and greater, we do not have an explicit understanding of s_n or v_n, in the sense that there is no (non-recursive) formula for the value of either s_n or v_n at a given location.

The configuration s_n is the fractal configuration. The odometer function v_n is mounded; see Example 5.5.1. Although the structure of v_n seems considerably simpler than s_n, it is equally difficult to understand at an exact level – note that the two uniquely determine each other.

In comparison, in the one-dimensional case, $\mathrm{stab}(\delta_0(n))$ is one of two configurations depending only on the parity of n. Either a solid segment of chips or a solid segment with a gap at the origin and the odometer function is equal to the triangle numbers; see Theorem 5.3.2.

5.5.1 Odometer

There is an easy recursion for the odometer function. Let $v_{i,j}$ be the total number of times that location (i, j) fires in the stabilization of an initial configuration on $\mathbb{Z}^2$. Then

$$v_{i,j} = \left\lfloor \frac{v_{i-1,j}}{4} + \frac{v_{i,j-1}}{4} + \frac{v_{i+1,j}}{4} + \frac{v_{i,j+1}}{4} + \delta_{ij} \right\rfloor. \tag{5.2}$$

The next example gives a sense of the number of firings for the pulse in two dimensions.

Example 5.5.1. Below is the number of times each site fires in the stabilization of $\delta_0(188)$. The initial configuration guarantees that the center site will fire 47 times. But we see that it fires 86 times, hence the chips are fired back to the origin enough times to cause 39 additional firings. The 39 additional firings require a total of 156 chips to return to the origin or almost $\frac{5}{6}$th of the total number of chips.

		1	2	3	2	1		
	2	4	7	9	7	4	2	
1	4	9	15	19	15	9	4	1
2	7	15	27	39	27	15	7	2
3	9	19	39	86	39	19	9	3
2	7	15	27	39	27	15	7	2
1	4	9	15	19	15	9	4	1
	2	4	7	9	7	4	2	
		1	2	3	2	1		

As the example illustrates, chip-firing results in a lot of firings. There is however a sense in which this system is doing the least amount of work possible to reach the final configuration.

From the initial configuration $\delta_0(n)$, the chip-firing process produces a unique final configuration s_n. In turn, this uniquely defines the odometer v_n. However, there are other solution pairs to the Laplace equation 5.3 which are not given by the chip-firing process. Namely, there exist pairs (s'_n, v'_n) such that

$$s'_n = \delta_0(n) + \Delta v'_n. \tag{5.3}$$

The key is that s'_n is not the result of *chip-firing* from $\delta_0(n)$. Here again we have the *least action principle*; see Lemma 2.6.18.

Theorem 5.5.2. *Of all solutions to Equation 5.3 the solution given by chip-firing, (s_n, v_n), is minimal in the following sense: for any other solution (s'_n, v'_n), $v_n \leq v'_n$ coordinate-wise.*

Let us interpret this result. Even if we allowed for solutions reachable by illegal firings or cluster firings, or any other mechanism, Theorem 5.5.2 states that in some sense the usual chip-firing process actually does the least amount of work over all other solutions.

5.5.2   Support

What are some of the simplest things we can observe about s_n? First, we can consider the support of $\delta_0(n)$, which by abuse we will also refer to as the support of $s_n (= \text{stab}(\delta_0(n)))$. Recall that the support of an initial configuration **c** is defined to be the collection of all sites that fire during the chip-firing process started at **c**. The support appears to be approximately circular. Experimental inspection suggests that it is not exactly circular – at the boundary, along each axis, for example, is an observable flat side of the support. Again, as a reflection of the difficulty of these questions, it is an open problem to prove that the support is not converging to a circle. There are however a number of bounds on the support.

First, we show a convexity result of Le Borgne and Rossin [LBR02]. We establish the result for the first octant, $0 \leq j \leq i$, the rest of the plane follows by symmetry.

Similar to the recursion of Equation 5.2, define $c_{i,j}(t)$ to be the total number of chips that have been fired to site (i, j) in the first t time steps of the stabilization of $\delta_0(n)$. Then if $(i, j) \neq (0, 0)$,

$$c_{i,j}(t+1) = \lfloor \frac{c_{i-1,j}(t)}{4} \rfloor + \lfloor \frac{c_{i,j-1}(t)}{4} \rfloor + \lfloor \frac{c_{i+1,j}(t)}{4} \rfloor + \lfloor \frac{c_{i,j+1}(t)}{4} \rfloor. \tag{5.4}$$

For $(i,j) = (0,0)$, define $c_{0,0}(0) = n$. For $t > 0$, $c_{0,0}(t)$ follows the recurrence of Equation 5.4 with an additional summand of n.

The support consists of all those sites with $c_{i,j}(t) > 4$ for some value of t. The next proposition investigates the number of chips fired to sites along four different paths; see Figure 5.7.

Proposition 5.5.3. *The following four systems of sequences, with starting point in the first octant, are all decreasing:*

$$A : [c_{l,2i+m}(t)] \quad B : [c_{2i+m,l}(t)] \quad D : [c_{l+i,l-i}(t)] \quad E : [c_{l+i,i}(t)]$$

Proof. The proof proceeds by induction on t. The claim is directly checked for $t = 1$.

For the general case, consider two sites corresponding to adjacent terms in a sequence. The idea is to map the neighbors of one site to the neighbors of the previous site and then invoke Equation 5.4. The only potential difficulty lies at the diagonal. However, sites that are diagonal neighbors across the diagonal line $i = j$ are symmetric and have equal c values. □

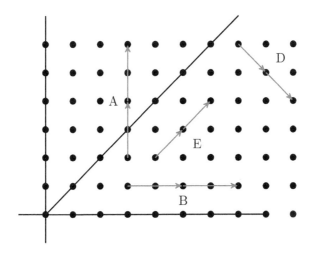

FIGURE 5.7: The decreasing sequences of Proposition 5.5.3.

Proposition 5.5.4. *For $i > j > 0$, if site (i,j) is in the support of $\delta_0(n)$ then site $(i-1,j)$ is also in the support of $\delta_0(n)$.*

Proof. Using the sequences of Proposition 5.5.3 we show that all neighbors of site $(i-1,j)$ must fire.

If site (i, j) is in the support,
then sequence "B" shows that $(i - 2, j)$ is in the support.

If site $(i - 2, j)$ is in the support,
then sequence "E" shows that $(i - 1, j + 1)$ is in the support.

If site $(i - 1, j + 1)$ is in the support,
then sequence "A" shows that $(i - 2, j + 1)$ is in the support.

Now all neighbors of site $(i - 1, j)$ have fired at some point; thus $(i - 1, j)$ must fire and is in the support. □

Using this convexity, Le Borgne and Rossin prove that the support is bounded by adjacent L_1 and L_∞ balls. Let

$$L_1(r) = \{(x, y) \mid |x| + |y| \leq r\} \text{ and } L_\infty(r) = \{(x, y) \mid |x| \leq r, |y| \leq r\}.$$

Theorem 5.5.5 ([LBR02]). *Let S_n denote the support of $\delta_0(n)$. Then for some r,*

$$L_1(r - 1) \subseteq S_n \subseteq L_\infty(r).$$

Proof. 1. Let r be the maximum over all coordinates of all sites in the support S_n. Then by definition, $S_n \subseteq L_\infty(r)$.

2. Let (i, j) be a site in S_n such that $i = r$. Such a site is guaranteed to exist by the definition of r. Sequence A implies that sites $(r, j - 2k)$ with $j - 2k \geq 0$ are in S_n. Sequence E then implies that sites $(r - 1, j - (2k - 1))$ with $j - (2k - 1) \geq 0$ are in S_n. Proposition 5.5.2 implies that all sites to the left of these two sets of sites (in the first octant) are in S_n. All together S_n contains the L_1 ball of radius either r, if j is even or $r - 1$, if j is odd. □

For any subset W of vertices in the support of s_n, if W consists of k edges, then W must have at least k chips in s_n; see the proof of Proposition 5.2.1. This fact combined with Theorem 5.5.5 gives the following corollary.

Corollary 5.5.6. *The value of r in Theorem 5.5.5 is bounded as*

$$\frac{\sqrt{n} - 1}{2} \leq r \leq \frac{\sqrt{n}}{2}.$$

Additionally, Landau, Levine and Peres bound the support of s_n between two L_2 balls. Let B_R denote the L_2 ball of radius R.

Theorem 5.5.7 ([LLP08] [LP09]). *Let S_n denote the support of $\delta_0(n)$. Then there exist R_1 and R_2 such that,*

$$B_{R_1} \subset S_n \subset B_{R_2}.$$

For $n = \pi r^2$, $R_1 \geq r/\sqrt{3}$ and $R_2 \leq (r + o(r))/\sqrt{2}$.

5.5.3 Backgrounds

As already suggested, it appears as though the boundary of S_n is not approaching a circle as n gets large. The boundary might be more appropriately described as polygonal. Next we consider S_n in the context of a family of configurations on the grid whose supports mediate between a square and a circle.

Consider the initial configuration with n chips at the origin and x chips at every other location, where $x \leq 2$ (including negative values). The value x is thought of as the *background height*. The discrete pulse of the last few sections is the case $x = 0$. We do not consider $x \geq 3$ because these initial configurations (a pulse with background height $x \geq 3$) do not stabilize.

Figures 5.8–5.12 show the stabilization of the pulse for background heights $x = 2, 1, 0, -1, -2$.

Clearly, the support of the final configuration changes with different background height. The precise shape of the support can be determined for the maximum stabilizing background height.

Proposition 5.5.8. *The support of the stabilization of the pulse $\delta_0(n)$ with background height $x = 2$ is a perfect square.*

Let r_x be the radius of the support of the pulse with background height x. Le Borgne and Rossin [LBR02] showed that for $x = 0, 1, 2$, r_x is $\Theta(\sqrt{n})$ and experimentally:

$$r_0 \sim .383\sqrt{n},$$
$$r_1 \sim .484\sqrt{n},$$
$$r_2 \sim .800\sqrt{n}.$$

Furthermore, Fey, Levine and Peres [FLP10] have shown that for all x, r_x is $\Theta(\sqrt{n})$.

As seen in Figures 5.8–5.12, as the background decreases, the support of the final configuration becomes more and more circular. As x goes strongly negative, the support of the pulse, as n grows, is in fact very close to a circle; see [LP09].

FIGURE 5.8: The stabilization of the pulse with background height 2.

FIGURE 5.9: The stabilization of the pulse with background height 1.

FIGURE 5.10: The stabilization of the pulse with background height 0.

FIGURE 5.11: The stabilization of the pulse with background height -1.

FIGURE 5.12: The stabilization of the pulse with background height −2.

5.5.3.1 Higher dimensions

In the last few sections, we have been discussing the two-dimensional case. One can easily imagine a discrete pulse at the origin of the $\mathbb{Z}^d$ lattice for $d \geq 3$. Proposition 5.5.8 and the asymptotic result $r_x \in \Theta(n^{\frac{1}{d}})$ extend to arbitrary dimension. Experimentally, it is of course harder to visualize the resulting patterns. For the background heights $x = 2d - 2$, the support is a cube. For $0 \leq x \leq 2d - 2$ the limiting region is (smoothed) polytopal. Figure 5.13 shows two-dimensional slices of the support of the pulse in three dimensions with no background, i.e. $x = 0$.

5.5.4 Scaling limits

Only recently has there been success proving some of the observed behavior seen in simulation. Here we describe the results of Pegden and Smart [PS13] which proves the existence of a type of *scaling limit*. The result does not explain the intricate patterns, but does confirm that the patterns are in a sense converging as n grows.

Once again we are interested in

$$s_n = \text{stab}(\delta_0(n)) = \delta_0(n) + \Delta v_n. \tag{5.5}$$

Next we need to introduce scaling factors. Why do we consider a scaling limit? Note that as n increases neither s_n nor v_n are in fact converging. For example, the support of s_n increases to larger and large circular regions as n grows. The odometer function also increases – the more initial chips at the origin, the more times the origin will fire during stabilization. In Figures 5.1 and 5.2, each pattern is actually of a very different size. We have scaled the images to compare them. Visually, the limiting behavior manifests as a sharpening of the image – it seems we are witnessing higher and higher resolution images of an underlying limit.

We make the following definitions:

$$h = n^{\frac{-1}{d}},$$

$$\bar{s}_n(x) = s_n\left(\frac{1}{h}x\right),$$

$$\bar{v}_n(x) = n^{\frac{-2}{d}} v_n\left(\frac{1}{h}x\right).$$

Given a function $u_n : h\mathbb{Z}^n \to \mathbb{R}$ extend it to all of $\mathbb{R}^d$ by nearest neighbor interpolation: for $x \in \mathbb{R}^d$ define $u_n(x)$ to be the value of u_n at the nearest integer point (denoted by brackets): $u_n(x) = u_n(h[h^{-1}x])$.

FIGURE 5.13: Slices of the three-dimensional configuration formed from starting with 10 million chips at the origin of $\mathbb{Z}^3$.

Finally, the scaled Laplacian is

$$\Delta^h u(x) = \frac{1}{h^2} \sum_{y \sim x} (u(y) - u(x)).$$

The *fundamental solution* Φ_n of the scaled Laplacian is the function such that

$$\Delta^h \Phi_n(x) = \begin{cases} -n & \text{if } x = 0 \\ 0 & \text{otherwise.} \end{cases}$$

In order to see $\bar{s}_n$ directly as an image of the Laplacian, we define:

$$\bar{w}_n = \bar{v}_n - \Phi_n,$$

so that

$$\bar{s}_n = \Delta^h \bar{w}_n.$$

The sequence $\bar{w}_n$ converges locally uniformly but the nature of $\bar{s}_n$ requires a different form of convergence.

Definition 5.5.9. A sequence of functions $\bar{s}_n \in L^\infty(\mathbb{R}^d)$ converges weakly-* to a function $s \in L^\infty(\mathbb{R}^d)$ if

$$\int_{\mathbb{R}^d} \bar{s}_n \phi \, dx \to \int_{\mathbb{R}^d} s \phi \, dx \quad \text{as } n \to \infty$$

for all test functions $\phi \in C_0(\mathbb{R}^d)$.

A close examination of s_n for large n and the seeming fractal nature of s_n both suggest the need for a non-pointwise convergence. For example, Figure 5.14 shows a zoomed-in corner of s_n. Experiments show that the checkerboard pattern oscillates as n grows. Since the weak-* property gives convergence of local averages, informally, it can smooth out these oscillations.

The main result of Pegden and Smart is the following existence theorem.

Theorem 5.5.10 ([PS13]). *There are functions* $w \in C(\mathbb{R}^d)$ *and* $s \in L^\infty(\mathbb{R}^d)$ *such that* $\bar{s}_n$ *converges weakly-* *to* s *in* $L^\infty(\mathbb{R}^d)$, $\bar{w}_n$ *converges locally uniformly to* w *in* $C(\mathbb{R}^d)$, *and* w *is a weak solution of* $\Delta w = s$ *in* $\mathbb{R}^d$.

The proof of Theorem 5.5.10 uses the theory of partial differential equations. The *sandpile PDE* is further investigated in [LPS16], where solutions to the PDE are related to fractal structures called Apollonian triangulations.

While we could not hope for a pointwise convergence, weak-* convergence is still a relatively weak form of convergence. It allows for the oscillations seen in Figure 5.14 but also does not discern such patterns.

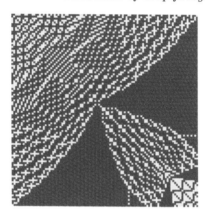

FIGURE 5.14: A zoomed-in corner of the configuration s_n.

Question 5.5.11. *Can one strengthen the form of convergence to something stronger than weak-*? What is the rate of convergence?*

5.6 Other lattices

One might ask how much of the interesting pattern formation depends on the exact structure of the grid graph. Here we show three new examples of fractal patterns. These three are on infinite planar lattices. In the next section, we will find similar patterns in the finite case.

Figure 5.15 shows the stabilization of a pulse on the planar lattice of degree 8 consisting of the grid and all 45 and −45 degree diagonals.

The next example is on the F-lattice. Every lattice point in the F-lattice has indegree 2 and outdegree 2; the lattice is shown in Figure 5.16. Chip-firing on directed graphs is defined in Chapter 6. The basic dynamics are as expected, a site fires if it has at least as many chips as its outdegree. Figure 5.17 shows the stabilization of the pulse on the F-lattice with a checkerboard background configuration; every other lattice location is started at height 1.

The last example is a pulse on the usual grid graph, but with a checkerboard background as in the case of the F-lattice. Figure 5.18 shows the stabilization of the pulse on the grid with a checkerboard pattern of 3s and 1s as background. This example should be compared to the pulse with background height uniformly equal to 2 as shown in Figure 5.8.

FIGURE 5.15: The stabilization of the pulse of 10 million chips on the planar lattice of degree 8 consisting of the grid and all 45 and −45 degree diagonals.

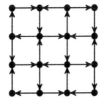

FIGURE 5.16: A patch of the F-lattice.

FIGURE 5.17: The stabilization of the pulse of 1 million chips on the F-lattice with checkerboard background 0/1.

FIGURE 5.18: The stabilization of the pulse of 1 million chips on the grid with checkerboard background 1/3.

5.7   The identity element

Thus far we have considered pattern formation on infinite graphs. Pattern formation has also been observed in the context of finite graphs, in particular, for the identity element of the sandpile group.

Recall that for a finite graph G with a sink, the critical configurations of G formed a distinguished system of representatives of the sandpile group $\mathcal{S}(G)$; see Chapter 4 Section 2.6.1. Addition in the sandpile group can be interpreted as superimposing two critical configurations and toppling the result until reaching a unique final critical configuration. For critical configurations $\mathbf{c}$ and $\mathbf{d}$, their sandpile sum is

$$\text{stab}(\mathbf{c} + \mathbf{d}).$$

Every group has an identity element and the firing-dynamic interpretation of the sandpile group leads one to picture the all zeros configuration as the identity element. Clearly $\text{stab}(\mathbf{c}+\mathbf{0}) = \mathbf{c}$. The problem is that the all zeros configuration is typically not a critical configuration. The all zeros configuration is instead always superstable – certainly there are no legal cluster fires. Criticality would imply that there is some *sufficiently large* initial configuration that can result in all chips eventually reaching the sink through a sequence of legal fires. This is rarely the case for the all zeros configuration.

How does one compute the identity element? Recall that there exists a unique maximal critical configuration $\mathbf{c}_{\max}$ which is equal to the degree sequence minus one; each vertex has as many chips as possible without being ready to fire. Further recall the following proposition from Chapter 4.

Proposition 5.7.1. *For a graph G, let e be the identity element of the sandpile group $\mathcal{S}(G)$. The identity element e can be computed from $\mathbf{c}_{\max}$ as follows:*

$$e = \text{stab}(2\,\mathbf{c}_{\max} - \text{stab}(2\,\mathbf{c}_{\max})).$$

Example 5.7.2. Consider the graph whose non-sink vertices form an $n \times n$ square grid and with a sink connected to all nodes on the boundary so that each vertex has degree four. Configurations on this graph can be thought of as sandpiles on a table top – if boundary sites fire, sand simply falls to the floor. Figures 5.19, 5.20 and 5.21 show the identity configuration for $n = 100$, 500 and 1000. See also Chapter 4 Section 4.3.

FIGURE 5.19: Grid graph identity element of the 100×100 grid.

FIGURE 5.20: Grid graph identity element of the 500×500 grid.

FIGURE 5.21: Grid graph identity element of the 1000×1000 grid.

Once again we see compelling patterns. Perhaps most stunning is that these configurations act as the identity. If we start with any critical configuration **c** on the grid and superimpose the configuration **c** with the configuration from Figure 5.21 then after stabilizing we return to the configuration **c**.

The most noticeable difference from the pulse is the large square of constant value around the origin. Much like the circularity of the support, we have not yet proved the existence of the square.

Conjecture 5.7.3 ([LBR02]). *The identity element of the $2n \times 2n$ grid has a central square of exactly two chips whose size is linear in the size of the grid. The identity element of the $2n + 1 \times 2n + 1$ grid has a central square consisting of two chips except for a single vertical and horizontal central line.*

More progress has been made for the identity elements of other lattices. The F-lattice, in particular, has been well studied by Caracciolo, Paoletti and Sportiello [CPS08]. Paoletti's book [Pao14] contains much more on the theory of pattern formation than is covered here. We recommend [Pao14] to the interested reader.

5.8 Exercises

Exercise 5.8.1. *Simulate chip-firing patterns. Some possibilities,*

1. *One-dimensional chip-firing from an initial configuration other than the pulse, e.g. two adjacent and equal sized stacks of chips.*

2. *Two-dimensional chip-firing of a pulse on other planar lattices such as the honeycomb.*

3. *Two-dimensional chip-firing from an initial configuration other than the pulse.*

Exercise 5.8.2. *Compute the odometer function for $n = 10, 100, 1000$ in each of the following cases:*

- *The pulse $\delta_0(n)$ of the two-dimensional grid.*

- *The pulse $\delta_0(n)$ of the two-dimensional grid with a 0/1 checkerboard background.*

- *The pulse $\delta_0(n)$ on the F-lattice.*

Exercise 5.8.3. *For labeled chip-firing with $n = 5, 7, 9$ chips, how many different orderings are possible in the final configuration?*

Exercise 5.8.4. *For labeled chip-firing with an odd number of chips, what is the maximum number of inversions of the final configuration?*

Exercise 5.8.5. *Suppose you have the labeled configuration: $(n, n-1, n-2, \ldots, 1)$ (chips are in adjacent locations along the one-dimensional grid). How many chip-firing and opposite chip-firing moves do you need to sort the chips into increasing order?*

Exercise 5.8.6. *Find the identity element of the sandpile group of the cycle graph C_n.*

Exercise 5.8.7. *Find the identity element of the wheel graph (with sink at either the center or along the boundary).*

Exercise 5.8.8. *The effect of the sink on the identity.*

1. *Take any finite planar graph and add a sink along the boundary. Compute the identity element.*

2. *Take any finite planar graph and add a sink vertex connected to every site. Compute the identity element.*

Exercise 5.8.9. *Prove Proposition 5.7.1 giving a formula for the identity element of the sandpile group.*

Exercise 5.8.10. *Let G be an $n \times n$ grid. Define β to be the configuration obtained from the all zeros configuration and reverse firing all non-sink sites. Define a sequence of configurations:*

$$\mathbf{d}^k = \text{stab}(\mathbf{d}^{k-1} + \beta).$$

Prove that the identity element is the configuration $\mathbf{d}^m$ such that

$$\mathbf{d}^m = \mathbf{d}^{m+1}.$$

Exercise 5.8.11. *Prove Proposition 5.5.8: the support of the pulse with background height 2 is a perfect square.*

Exercise 5.8.12. *Fill in the details of the proof of Proposition 5.5.3: that the four sequences $A - E$ are decreasing.*

Exercise 5.8.13. *(Open)*
Prove that the identity element of the grid graph has a large square of constant value centered at the origin.

Exercise 5.8.14. *(Open)*
Prove that the support of the pulse $\delta_0(n)$ is not approaching a circle as $n \to \infty$.

Exercise 5.8.15. *(Open)*
Prove that labeled chip-firing with an odd number of chips results in a sorted final configuration with probability approaching $\frac{1}{3}$ as $n \to \infty$.

Part II

Extensions

Chapter 6

Avalanche Finite Systems

Thus far we have primarily considered finite undirected graphs. In this chapter we broaden the context of chip-firing to more general networks. Our starting point is the graph Laplacian. .

In the context of finite graphs, the graph Laplacian is the operator which completely dictates the dynamics of firing. We will see that it is possible to extend the class of operators for chip-firing from graph Laplacians to more general matrices and still maintain much of the good behavior of graphical chip-firing.

In terms of generalized networks, a natural next step past undirected graphs would be directed graphs. We will set up an even more general chip-firing process and see directed graphs as a special case in Section 6.5.1.

6.1 M-matrices

Following the work of Dhar, Gabrielov [Gab94], [Gab93] considered more general chip-firing dynamics in terms of a class of dissipation matrices which is broader than graph Laplacians. Gabrielov called these *avalanche-finite matrices*. As the name suggests, they are precisely the matrices necessary so that all initial configurations eventually stabilize using legal firing moves dictated by the matrix. In Chapter 2 we saw that for a graph with a sink, all initial configurations stabilize. Hence for all finite undirected graphs, the reduced graph Laplacians are avalanche finite matrices.

Lorenzini [Lor89] worked with avalanche finite matrices in his study of *arithmetical graphs*. Postnikov and Shapiro [PS04] investigated avalanche finite matrices under the name *toppling matrices*. Avalanche finite matrices have also been studied in great detail in fields such as economics, operations research, finite difference and finite element analysis. In such contexts, they are known as *M-matrices*.

Before defining M-matrices, we set up the dynamics of the chip-firing process defined by an arbitrary matrix N. Suppose that we have an abstract network of sites and a commodity that is traded in some manner by a local pairwise rule. For an $n \times n$ matrix N, we consider a system with n sites. A *chip configuration* is any integer vector $\mathbf{c} \in \mathbb{Z}_{\geq 0}^n$. Firing a site i is defined to be the process which replaces the configuration $\mathbf{c}$ with $\mathbf{c} - N^T e_i$, i.e. subtracting the ith row of N from $\mathbf{c}$. Note that we now need to work in terms of the transpose of N. For finite undirected graphs the Laplacian is symmetric and so this distinction is unnecessary. A *legal* chip-firing move is one in which the resulting configuration $\mathbf{c} - N^T e_i$ is also non-negative. A site i is *ready to fire* if $c_i \geq N_{ii}$. A configuration is *stable* if $c_i < N_{ii}$ for all i.

A natural question arises: for which matrices does the chip-firing process always eventually stabilize versus those matrices which possibly produce an infinite process.

Definition 6.1.1. A non-singular matrix N is *avalanche-finite* if every non-negative chip configuration eventually stabilizes in the chip-firing process defined by N.

Following the physicality of the original model, Gabrielov restricted to matrices with a positive diagonal and non-positive off-diagonal. Therefore, a site must have a certain positive amount of chips in order to be ready to fire and firing a site increases the number of chips on neighboring sites.

Definition 6.1.2. A matrix L is a *Z-matrix* if the diagonal entries of L are non-negative and the off-diagonal entries are non-positive.

Z-matrices are referred to as *redistribution matrices* in [Gab94].

Theorem 6.1.3. *Let L be a $n \times n$ Z-matrix. If any of the following equivalent conditions hold then L is called a* non-singular *M-matrix:*

1. *L is avalanche finite.*

2. *L^T is avalanche finite.*

3. *The real part of all eigenvalues of L are positive.*

4. *L^{-1} exists and all the entries of L^{-1} are non-negative.*

5. *There exists a vector $x \in \mathbb{R}^n$ with $x \geq 0$ such that Lx has all positive entries.*

6. *There exists a vector $x \in \mathbb{R}^n$ with $x > 0$ such that Lx has all positive entries.*

7. *All principal minors of L are positive.*

8. *There exists a positive diagonal matrix D with $DL+(DL)^T$ positive definite.*

There are many more known equivalent conditions for M-matrices. Plemmons [Ple77] lists 40 different characterizations of non-singular M-matrices.

The equivalence of the first condition is due to Gabrielov [Gab94]. Briefly, suppose that condition 6 holds. Let $\mathbf{c}$ be a chip configuration for L. Then $x \cdot \mathbf{c}$ is non-negative, remains non-negative with each legal firing, and strictly decreases with each legal firing. Therefore there can only be finitely many legal firings.

Example 6.1.4. For a finite undirected graph G with sink q, the reduced Laplacian Δ_q is a non-singular M-matrix. Indeed, by Proposition 2.5.2, every initial configuration on a graph with a sink eventually stabilizes hence Δ_q is avalanche finite.

Alternatively, condition 6 is satisfied by Lemma 2.6.7 which constructs a non-negative vector $\mathbf{z}$ such that $\Delta_q \mathbf{z} > 0$ by looking at the row sums of the inverse of Δ_q.

Example 6.1.5. The following 2×2 matrix L is easily checked to be an M-matrix. On the other hand, L is not the reduced Laplacian of any graph, even allowing for directed, multiple edges.

$$L = \begin{pmatrix} 3 & -4 \\ -1 & 2 \end{pmatrix}.$$

Although L is not a graph Laplacian, we can still envision a two-site system.

- When site one fires, it loses 3 chips and site two gains 4 chips.

- When site two fires, it loses 2 chips and site one gains 1 chip.

In order to explain the overall loss of 1 chip when site two fires, we could envision a virtual sink. This does not work, however, to explain the overall gain of 1 chip when site one fires.

The asymmetry of the firing (site one fires 4 chips to site two but site two fires only 1 chip to site one), can be thought of as a "currency exchange." The 4 chips from site one are worth 1 chip from site two. Figure 6.1 shows an attempt at a graphical representation of the chip-firing network dynamics induced by the matrix L.

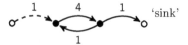

FIGURE 6.1: Representation of a network with a virtual "sink" and virtual "source." A sink vertex would naturally compensate for the loss of 2 chips at site two. A virtual source is less natural. The influx of 1 chip occurs when site one fires. These dynamics are not well modeled by a graphical network.

6.2   Chip-firing on M-matrices

In the chip-firing process with respect to an M-matrix, any initial configuration will stabilize – this is the defining principle we seek to maintain – but do we maintain more of the nice behavior of the chip-firing process?

Clearly confluence is maintained. If two sites can fire, firing either site still allows the other site to fire. This is due to the fact that firing any one site can only increase the number of chips at other sites, i.e. the off-diagonal entries are non-positive.

Additionally, there is a well-defined group structure on chip configurations. Define two configurations $\mathbf{c}$ and $\mathbf{d}$ in $\mathbb{Z}^n$ to be *firing-equivalent* under L if their difference $\mathbf{c} - \mathbf{d}$ is in the $\mathbb{Z}$-image of L:

$$\mathbf{c} - \mathbf{d} = L^T \mathbf{z}$$

for some $\mathbf{z} \in \mathbb{Z}^n$.

Definition 6.2.1. The *sandpile group* $\mathcal{S}(L)$ of an avalanche-finite matrix L is the integer cokernel of the transpose of L,

$$\mathcal{S}(L) = \mathbb{Z}^n \,/\, \mathrm{im}(L^T) = \mathrm{coker}_{\mathbb{Z}}(L^T).$$

Note furthermore that the size of the sandpile group $|\mathcal{S}(L)|$ is equal to the determinant $\det(L) = \det(L^T)$.

6.3 Stability

For a finite graph, the sandpile group can alternatively be defined over critical configurations. The notions of criticality and superstability can be extended to the context of M-matrices with only a few changes. Criticality will in fact stay the same, but we will see a subtle distinction in the definition of superstable configurations.

6.3.1 Superstability

Consider the following three notions of stable configurations, each strictly stronger than the previous. In each, the matrix L is an M-matrix.

Definition 6.3.1. A vector $\mathbf{c} \in \mathbb{Z}^n$ is *stable* with respect to L if for all i,

$$c_i < L_{ii}^T.$$

Definition 6.3.2. A vector $\mathbf{c} \in \mathbb{Z}^n$ with $\mathbf{c} \geq 0$ is *χ-superstable* if for every $\chi \in \{0, 1\}^n$ with $\chi \neq 0$ there exists $1 \leq i \leq n$ such that

$$c_i - (L^T \chi)_i < 0.$$

Definition 6.3.3. A vector $\mathbf{c} \in \mathbb{Z}^n$ with $\mathbf{c} \geq 0$ is *z-superstable* if for every $z \in \mathbb{Z}^n$ with $z \geq 0$ and $z \neq 0$ there exists $1 \leq i \leq n$ such that

$$c_i - (L^T z)_i < 0.$$

In Chapter 2 where we were only considering finite *undirected graphs*, we used χ-superstable as our definition of superstable configurations. The χ-superstable configurations are those configurations from which you cannot legally cluster-fire a collection of vertices (simultaneously fire multiple sites). The z-superstable configurations are those configurations from which you cannot legally cluster-fire with multiplicity (simultaneously fire multiple sites, some more than once).

In the case of finite graphs (and directed Eulerian graphs, see Section 6.5.1) this distinction is not necessary; the χ-superstable configurations are the same as the z-superstable configurations.

For general M-matrices, the two notions are not the same, one needs the extended notion of z-superstable configurations; see Example 6.3.4.

Example 6.3.4. Consider again the M-matrix L:

$$L = \begin{pmatrix} 3 & -4 \\ -1 & 2 \end{pmatrix}.$$

The $\det(L) = 2 = |\operatorname{coker}(L^T)|$. An explicit calculation shows the image of the three non-zero characteristic vectors:

$$L\begin{pmatrix} 1 \\ 0 \end{pmatrix} = \begin{pmatrix} 3 \\ -1 \end{pmatrix}, \quad L\begin{pmatrix} 0 \\ 1 \end{pmatrix} = \begin{pmatrix} -4 \\ 2 \end{pmatrix}, \quad L\begin{pmatrix} 1 \\ 1 \end{pmatrix} = \begin{pmatrix} -1 \\ 1 \end{pmatrix}.$$

Hence, for a configuration to be χ-superstable, the first coordinate must be strictly less than 3 and the second coordinate must be strictly less than 1. The χ-superstable configurations are:

$$(2,0), (1,0), \text{ and } (0,0).$$

Of these three configurations, $(2,0)$ is not z-superstable since it is in the $\mathbb{Z}$-image of L,

$$L\begin{pmatrix} 2 \\ 1 \end{pmatrix} = \begin{pmatrix} 2 \\ 0 \end{pmatrix}.$$

Note this shows that $(2,0)$ and $(0,0)$ are equivalent under L and so the χ-superstable configurations are not unique per firing equivalence class induced by L.

The z-superstable configurations for L are

$$(1,0) \text{ and } (0,0).$$

There are two z-superstable configurations in total and their difference is not in the $\mathbb{Z}$-image of L. So, the z-superstables do form a system of distinct representatives for the firing equivalence classes of L. We will see shortly that this is always the case.

6.3.2 Criticality

Critical configurations are defined exactly as in the graphical case: those stable configurations which are reachable from a sufficiently large initial configuration as a result of iterated legal moves.

Definition 6.3.5. A configuration $\mathbf{c} \in \mathbb{Z}^n$ is *critical* if $\mathbf{c}$ is

1. Stable.

2. Reachable from a sufficiently large initial configuration $\mathbf{b}$, where a configuration is sufficiently large if every site is ready to fire: $\mathbf{b}_i > L_{ii}$ for all i.

Most of the equivalences of critical configurations from the graphical case, see Theorem 2.6.3, extend to the setting of M-matrices.

In order to state the equivalences, we first need the following two definitions. For an integer vector $\mathbf{b} = (b_1, b_2, \ldots, b_n)$, define the *support of* **b** supp(**b**) as the components of **b** that are non-zero: supp(**b**) = $\{i \; b_i \neq 0\}$. The *support graph* of an $n \times n$ M-matrix is the directed graph on n vertices with an edge directed from vertex i to vertex j whenever $L_{ij} < 0$. Namely, there is a directed edge from i to j if firing site i causes site j to gain chips.

Theorem 6.3.6 (compare to Theorem 2.6.3). *Let L be an $n \times n$ M-matrix. The following are equivalent for configurations* **c** *over L.*

1. **c** *is critical:* $\mathbf{c} = \mathrm{stab}(\mathbf{b})$ *for some* **b** *with $b_i \geq L_{ii}$ for all i.*

2. $\mathbf{c} = \mathrm{stab}(\mathbf{c} + \mathbf{b})$ *for some* **b** *with $b_i > 0$ for all i.*

3. $\mathbf{c} = \mathrm{stab}(\mathbf{c} + N\mathbf{b})$ *for some* **b** *with $b_i > 0$ and every integer $N \geq 1$.*

4. $\mathbf{c} = \mathrm{stab}(\mathbf{c} + \mathbf{b})$ *for some* **b** *such that for for each site j, there exists a directed path from i to j in the support graph of L from at least one site i in* supp(**b**).

6.3.3 Energy minimization

Given an M-matrix L and an integer vector **c** define the following energy form,

$$E(\mathbf{c}) = \|L^{-1}\mathbf{c}\|_2^2,$$

where $\|v\|_2^2 = v \cdot v$.

Given $\mathbf{c} \in \mathbb{Z}^n$ with $\mathbf{c} \geq 0$ consider the optimization:

$$\underset{\mathbf{d} \sim \mathbf{c}, \mathbf{d} \geq 0}{\mathrm{argmin}} \; E(\mathbf{d}).$$

For an M-matrix L, a solution to the minimization problem is a non-negative configuration equivalent to **c** with the smallest possible energy. We call such a configuration an *energy-minimizer* with respect to L.

Example 6.3.7. Again consider the M-matrix

$$L = \begin{pmatrix} 3 & -1 \\ -4 & 2 \end{pmatrix}.$$

The inverse of L is:

$$L^{-1} = \begin{pmatrix} 1 & .5 \\ 2 & 1.5 \end{pmatrix}.$$

Let **c** be the configuration $(4, 4)$. Then the energy of **c** is computed as:

$$E(L^{-1}\mathbf{c}^T) = E((6, 14)) = 6^2 + 14^2 = 232.$$

Firing site 1 yields a new configuration $\mathbf{c}' = (1, 8)$. The energy of $\mathbf{c}'$ is:

$$E(L^{-1}\mathbf{c}^T) = E((5, 14)) = 5^2 + 14^2 = 221.$$

As seen in Example 6.4.4, single firing moves eventually yield the critical configuration $(2, 0)$. After these firings, the energy has decreased to:

$$E(L^{-1}\mathbf{c}^T) = E((2, 4)) = 2^2 + 4^2 = 20.$$

The energy-minimizers are not the critical but the superstable configurations. Therefore in order to find the energy minimizer, we must continue to cluster-fire until reaching superstability. In this case, the equivalent superstable configuration is $(0, 0)$ which is clearly the energy-minimizer of its equivalence class.

6.3.4 Uniqueness

The configuration $\mathbf{c}_{\max}$ for an M-matrix L is equal to the diagonal of L minus one in each coordinate.

$$\mathbf{c}_{\max}(L) = \text{diag}(L) - \mathbf{1}.$$

Theorem 6.3.8 ([GK15]). *Let L be an M-matrix.*

1. *For every firing equivalence class defined by L, there exists unique critical, z-superstable and energy minimizing configurations.*

2. *The collection of z-superstable configurations coincides with the collection of energy minimizers.*

3. *The critical and superstable configurations are dual to each other via $\mathbf{c}_{\max}$: a configuration $\mathbf{c}$ is critical if and only if $\mathbf{c}_{\max} - \mathbf{c}$ is z-superstable.*

The proofs of the claims of Theorem 6.3.8 follow precisely the same arguments as in Chapter 2. Indeed the presentation of these results for finite graphs in Chapter 2 follows the more general setting in [GK15].

6.4   Burning

Recall the concept of a burning configuration from Chapter 2 Section 2.6.7. Burning configurations allow one to check if a configuration is critical. In the finite undirected graphical case, firing the sink gave the minimal burning configuration to check for criticality; each non-sink vertex would fire exactly once in the restabilization. In the M-matrix setting, burning configurations also allow one to check for criticality.

Definition 6.4.1. Let L be an M-matrix. A configuration $\mathbf{b} \geq 0$ is a *burning configuration* for L if

(i) $\mathbf{b}$ is in the integer image of L, $\mathbf{b} = L^T \mathbf{z}$ for some integer vector $\mathbf{z}$.

(ii) For each site j, there exists a directed path from i to j in the support graph of L from at least one site i such that $b_i \neq 0$.

Compare the next proposition to Proposition 2.6.27.

Proposition 6.4.2. *Let* $\mathbf{b}$ *be a burning configuration for an avalanche-finite matrix* L. *A configuration* $\mathbf{c}$ *is critical with respect to* L *if and only if* $\mathrm{stab}(\mathbf{c} + \mathbf{b}) = \mathbf{c}$.

Note also that Proposition 6.4.2 is Equivalence 4 of Theorem 6.3.6.

Definition 6.4.3. Given a burning configuration $\mathbf{b}$ for L, the *burning script* for $\mathbf{b}$ is

$$\mathbf{z} = (L^T)^{-1}\mathbf{b}.$$

The burning script records how many times sites fire when using a burning configuration. The stabilization of $(\mathbf{c} + \mathbf{b})$ has length $\sum_i z_i$ with site i firing precisely z_i times.

Example 6.4.4. Again consider the M-matrix:

$$L = \begin{pmatrix} 3 & -4 \\ -1 & 2 \end{pmatrix}.$$

The z-superstable configurations for L are

$$(0, 1) \text{ and } (0, 0).$$

The $\mathbf{c}_{\max}$ configuration is $(2, 1)$, therefore the critical configurations are

$$(2, 0) \text{ and } (2, 1).$$

Therefore, for any sufficiently large initial state, we must arrive either at $(2,0)$ or $(2,1)$.

Suppose that we wanted to check whether or not a configuration is critical. Given the configuration

$$\mathbf{c} = (2,0),$$

we could attempt to check if $\mathbf{c}$ is critical by demonstrating a sufficiently large configuration that stabilizes to $\mathbf{c}$.

For example let

$$\mathbf{a} = (4,4).$$

It is easily checked that $\text{stab}(\mathbf{a}) = \mathbf{c}$.

FIGURE 6.2: Stabilizing an initial configuration in order to verify criticality. A strike-through represents firing a vertex.

Theorem 6.4.2 gives an alternative way to check criticality via a burning configuration. Consider the configuration

$$\mathbf{b} = (1,0).$$

The configuration $\mathbf{b}$ is in the integer image of L:

$$L \begin{pmatrix} 1 \\ 2 \end{pmatrix} = \begin{pmatrix} 3 & -1 \\ -4 & 2 \end{pmatrix} \begin{pmatrix} 1 \\ 2 \end{pmatrix} = \begin{pmatrix} 1 \\ 0 \end{pmatrix}.$$

Since $\mathbf{b} \geq 0$ and $\mathbf{b} = L\mathbf{z}$ for some integer vector $\mathbf{z}$, $\mathbf{b}$ is a burning configuration for L.

$\text{stab}(\mathbf{c} + \mathbf{b})$ is equal to $\mathbf{c}$ as seen below:

$$(2,0) + (1,0) = (3,0).$$

Figures 6.2 and 6.3 are visualizations of stabilizations. Figure 6.2 shows the stabilization of the initial configuration $(4,4)$. A strikethrough represents a single firing. The stabilization of $(4,4)$ is the configuration $(2,0)$. Because the initial configuration was sufficiently large, this computation shows that $(2,0)$ is critical. Figure 6.3 shows the stabilization process using a burning configuration. The computation using the burning configuration is considerably shorter. A close inspection shows that

FIGURE 6.3: Using a burning configuration to see criticality.

the burning configuration allowed us to "jump ahead" in the first stabilization; $(\mathbf{c}+\mathbf{b}) = (3,0)$ is the fourth to last pair of values from Figure 6.2.

The vector $\begin{pmatrix} 1 \\ 2 \end{pmatrix} = (L^T)^{-1} \begin{pmatrix} 1 \\ 0 \end{pmatrix}$ is the burning script for $\mathbf{b}$. It records how many times each site fires in the stabilization of $(\mathbf{c} + \mathbf{b})$. Indeed site 1 fired once and site 2 fired twice.

Proposition 6.4.5 ([PPW13], see also [Spe93]). *For an M-matrix L, there exists a unique minimal burning configuration $\mathbf{b}$ in that for any other burning configuration $\mathbf{b}'$ for L,*

$$\mathbf{z_b} \leq \mathbf{z_{b'}},$$

where $\mathbf{z_b}$ is the burning script for $\mathbf{b}$ and similarly for $\mathbf{z_{b'}}$.

Perkinson, Perlman and Wilmes further describe how to construct the minimal burning configuration. For an M matrix L, let $\mathbf{b}$ equal the sum of the columns of L. If $\mathbf{b} \geq \mathbf{0}$, then $\mathbf{b}$ is the minimal burning configuration. Otherwise, if $\mathbf{b}_i < 0$ then update $\mathbf{b}$ by adding the ith column of L. Continue this process until $\mathbf{b} \geq \mathbf{0}$.

6.5 Directed graphs

In this section we will look more closely at chip-firing on directed graphs. There has been much work to extend properties of chip-firing to the directed graph case, e.g. [BL92], [Spe93], [Wag00], [HLM+08], [AB11], [PPW13]. Reduced graph Laplacians are one of the most important examples of M-matrices. The more general perspective of M-matrices helps to unify the various results and difficulties encountered for directed graphs.

6.5.1   Digraphs

Definition 6.5.1. Let G be a directed rooted graph with root vertex r. A subgraph T of G is an *arborescence rooted at r* if T contains all vertices of G and there is a unique directed path in T from every vertex to r.

Suppose that G is a directed multigraph on n vertices with a sink and at least one arborescence rooted at the sink. The chip-firing process on G is defined as follows: a vertex v is ready to fire if the number of chips at v is at least the outdegree of v. Firing a vertex v results in adding a chip to each vertex w such that $(v \to w)$ is a directed edge of G and subtracting $\mathrm{outdeg}(v)$ chips from v. At each time step, a vertex that is ready to fire is chosen and fired. The process terminates if at any time there are no sites with at least as many chips as outdegree. As before, firings can be described via a Laplacian.

Definition 6.5.2. Let G be a directed graph with n vertices. The *directed graph Laplacian* $\Delta(G)$ is an $n \times n$ matrix given by

$$\Delta_{ij} = \begin{cases} -1 & \text{if } i \neq j \text{ and } (v_i \to v_j) \in E, \\ \mathrm{outdeg}(v_i) & \text{if } i = j, \\ 0 & \text{otherwise.} \end{cases}$$

For a *directed multi-graph*, Δ_{ij} is equal to the number of oriented edges from v_i to v_j.

Therefore, firing is subtracting rows of a Laplacian. A configuration $\mathbf{c}'$ is obtained from a configuration $\mathbf{c}$ after firing site i if

$$\mathbf{c}' = \mathbf{c} - \Delta^T e_i.$$

An important subclass of directed graphs are Eulerian directed graphs.

Definition 6.5.3. A directed graph G is called *Eulerian* if for each vertex v in G, $\mathrm{outdeg}(v) = \mathrm{indeg}(v)$.

In terms of their Laplacians, the key feature of Eulerian graphs is that they have non-negative column sums.

Many extensions of undirected graph properties have been phrased in terms of Eulerian directed graphs. Since directed graph Laplacians are M-matrices, the previous section guarantees unique superstable and critical configurations along with the duality relation through the diagonal shifted by one when chip-firing on an Eulerian directed graph.

On the other hand, many properties which coincided for undirected graphs show subtle distinctions in the directed case. Eulerian graphs seem to be the most natural class to which many nice properties of the undirected case can be extended. For example, compare the next proposition to Theorem 3.6.3.

Definition 6.5.4. For a directed graph G on n non-sink vertices and sink vertex q, a G-parking function is a sequence of integers $(p_1, \ldots, p_n)$ such that for every nonempty subset $A \subseteq V \setminus q$, there exists a vertex $v \in A$ such that

$$p_v < \text{outdeg}_A(v),$$

where $\text{outdeg}_A(v)$ is the number of edges from v to vertices not contained in A.

Proposition 6.5.5. *For an Eulerian directed graph G, the following coincide:*

- χ*-superstable configurations,*

- z*-superstable configurations, and*

- G*-parking functions.*

We leave the proof as an exercise. The key technical fact is that in this case, the Laplacian has non-negative column sums.

In the non-Eulerian directed graph case, none of these collections coincide.

Proposition 6.5.6. *There exist non-Eulerian directed graphs G such that no two of the collections:*

$$\chi\text{-superstables, } z\text{-superstables, } G\text{-parking functions}$$

are equal.

Example 6.5.7 ([GK15]). Consider the directed graph G on 3 vertices with sink q and directed graph Laplacian equal to:

$$\begin{pmatrix} 3 & -3 & 0 \\ -1 & 2 & -1 \\ 0 & 0 & 0 \end{pmatrix}.$$

The graph G is illustrated in Figure 6.4. Vertex v_1 has three edges directed to vertex v_2. Vertex v_2 has a single directed edge to vertex v_1 and a single directed edge to the sink. The sink has no outgoing edges.

The transpose of the reduced graph Laplacian is:

$$L^T = \begin{pmatrix} 3 & -1 \\ -3 & 2 \end{pmatrix}.$$

FIGURE 6.4: A non-Eulerian directed graph.

It is not hard to check that all four 0/1-vectors of length two are χ-superstable for this graph. On the other hand, the all ones vector is not z-superstable as it is equal to $L(1,2)^T$. In particular, the all ones configuration is equivalent to the all zeros configuration.

For this graph, $\mathbf{c}_{\max} = (2,1)$.
The critical configurations are $(2,1)$, $(1,1)$, and $(2,0)$.
The z-superstables are $(0,0)$, $(1,0)$, $(0,1)$.
The G-parking functions are $(0,0)$, $(1,0)$, $(2,0)$; see the table below.

None of the collections are the same. To see, for example, that the z-superstable configuration $p = (0,1)$ is not a G-parking function, consider the subset $A = \{v_1, v_2\}$. We have that

$$\text{outdeg}_A(v_1) = 0$$

and

$$\text{outdeg}_A(v_2) = 1$$

and therefore p_i is not less than $\text{outdeg}_A(i)$ for any i.

χ-super	z-super	critical	G-parking
$(1,1)$	$(0,1)$	$(2,1)$	$(2,0)$
$(0,1)$	$(1,0)$	$(1,1)$	$(1,0)$
$(1,0)$	$(0,0)$	$(2,0)$	$(0,0)$
$(0,0)$			

The precise relationship between parking functions and chip-firing in the general case is considered and well discussed in [PS04]. We summarize here; see also [AB11].

For a directed graph $G = (V, E)$ with Laplacian Δ, a configuration $\mathbf{c}$ is *allowable* if for all subsets of vertices $I \subseteq V$ there exists a vertex $j \in I$ such that

$$c_j \geq \sum_{i \in I \setminus j} -\Delta_{ij}.$$

It is not difficult to show that every critical configuration is allowed.

Dhar conjectured that critical configurations might be characterized as those configurations that are both stable and allowed. (Recall that our first definition of a critical configuration was a configuration that is stable and reachable.) This conjecture is false in the most general case but true for M-matrices with non-negative column sums.

Proposition 6.5.8 ([Gab93])**.** *Let L be an M-matrix with non-negative column sums. Then a chip configuration for the chip-firing process determined by L is critical if and only if it is stable and allowed.*

Note that an M-matrix L is the transpose of the reduced Laplacian of some directed graph if and only if L has non-negative column sums.

Proposition 6.5.9 ([PS04])**.** *Let G be a directed graph with a sink and at least one arborescence. Let L be the transpose of the reduced Laplacian of G. A chip configuration for the chip-firing process determined by L is stable and allowed if and only if $\mathbf{c}_{\max} - \mathbf{c}$ is a G-parking function.*

In Example 6.5.7, the transpose of the reduced Laplacian has non-negative columns sums. The reduced Laplacian of G itself does not have non-negative column sums. Proposition 6.5.9 implies that the dual of a parking function is stable and allowed. The G-parking function $(2,0)$, for example, shows that Proposition 6.5.8 does not necessarily hold when L does not have non-negative column sums. Indeed $\mathbf{c}_{\max} - (2,0) = (0,1)$ is not a critical configuration for L.

Thus, parking functions are not necessarily the same as superstable configurations for general non-Eulerian directed graphs.

6.5.2 Stabilization

Non-singular M-matrices are precisely avalanche finite matrices. For graphs, these correspond to *reduced* graph Laplacians. For the next two sections, we pause our consideration of non-singular M-matrices in order to consider directed graphs without a sink vertex.

Theorem 2.3.6 gives three regimes of stabilization for chip configurations of undirected graphs in terms of the total number of chips in the initial configuration. There is not a similar result in the directed case, but there are partial results. Björner and Lovász explore the directed graph case in great detail in [BL92].

Definition 6.5.10. A directed graph G is called *strongly connected* if for every pair of vertices u, v in G, there is a directed path from u to v and from v to u.

Theorem 6.5.11 ([BL92]). *Let G be a strongly connected directed graph. Suppose the maximum size of any collection of edge-disjoint directed cycles is m. Then any initial configuration on G with less than m chips will stabilize.*

Proof. Consider chip-firing on G. Note when a chip first fires across an edge of a directed cycle. In all future steps of the chip-firing process, fire this chip only along this cycle. Because the number of chips is less than m, the vertices of some cycle will never fire. But, in a strongly connected graph, if the chip-firing process is infinite, then all vertices must fire infinitely often. $\square$

Proposition 6.5.12 ([Wag00] [BL92]). *Let G be a strongly connected graph with Laplacian Δ. Then there exists a unique vector $\mathbf{h} = (h_1, \ldots, h_n) \in \mathbb{Z}_{\geq 0}^n$ such that*

$$\Delta^T \mathbf{h} = 0 \text{ and } \gcd\{h_v\} = 1.$$

Moreover, $\ker(\Delta^T) = \mathbb{R}\mathbf{h}$.

Wagner refers to h_v as the activity of v. Here though activity is referring to the number of times a vertex v fires. We will reserve the term activity for the matroid context as in Section 3.2.3. For fully bidirected graphs, $\mathbf{h} = \mathbf{1}$.

Proposition 6.5.13 ([BL92]). *For a strongly connected graph G and a configuration $\mathbf{c}$, if there is a sequence of legal fires starting at $\mathbf{c}$ such that each site fires at least h_v times, then $\mathbf{c}$ does not stabilize.*

Proposition 6.5.13 parallels Lemma 2.3.2 from Chapter 2 where each site firing once in the chip-firing process with initial configuration $\mathbf{c}$ is enough to conclude that the process will never stabilize.

Farrell and Levine [FL16] consider using Proposition 6.5.13 to determine if a configuration $\mathbf{c}$ stabilizes: From a configuration $\mathbf{c}$, start chip-firing. Either a stable configuration will be reached or each site will fire enough times to invoke Proposition 6.5.13. Unfortunately, this method can take exponentially long as the next example shows.

Example 6.5.14 ([FL16]). Consider the directed multigraph consisting of a chain of directed cycles as shown in Figure 6.5. The edge labels represent edge multiplicities. Therefore, they also represent how many chips are fired to each neighbor. If there are a total of n vertices in this chain graph, then $h_i = 2^i 3^{n-i}$.

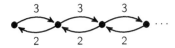

FIGURE 6.5: A graph for which **h** has exponentially large entries.

Farrell and Levine have given a simple criterion for checking whether or not a configuration will stabilize in the special case of *CoEulerian* graphs. Furthermore they show that determining whether or not an initial configuration will stabilize is NP-complete for general directed multigraphs.

The vector **h** is related to the periods of non-stabilizing chip-firing processes. Consider the chip-firing process in which at every time step, all sites that can legally fire do so. In this *parallel chip-firing* process, either a stable configuration is reached or the process becomes periodic. The entries of **h** provide a lower bound on the possible periods of non-stabilizing sequences. The periodic behavior of the parallel chip-firing process is studied in numerous sources including [BG92], [Pri94], [KNTG94], [GM97], [BCFV03] and [Lev11a].

6.5.3 Toppling time

An important difference in the theory of chip-firing on directed graphs is seen in the toppling time of those initial configurations that do stabilize. For an undirected graph (without a sink) we saw in Section 2.4 that if a configuration stabilizes, then the length of stabilization is polynomially bounded. Eriksson showed that this no longer holds for directed graphs.

Theorem 6.5.15 ([Eri91]). *For a directed graph on n vertices, there exist initial configurations which require exponentially (in n) many chip-firing moves before termination.*

Eriksson gives the following construction. Consider the directed wheel graph W_n with n vertices, where n is even, with all but one edge bidirected. Explicitly, construct a graph consisting of a cycle of length $n - 1$ and one additional vertex connected to every vertex of the cycle. Bidirect all edges of the outer cycle and all but one edge connected to the center vertex. For the last edge, direct it from the center vertex outwards to the cycle; see Figure 6.6.

The initial configuration consisting of $3n - 5$ chips on the center

FIGURE 6.6: The wheel graph with one directed edge from the center node. An initial configuration that takes exponentially many firing moves to stabilize consists of $3n - 5$ chips at the center vertex and no chips elsewhere.

vertex and 0 chips everywhere else stabilizes only after exponentially many chip-firing moves.

For W_n, the quantity $3n - 5$ is equal to $\sum \mathbf{c}_{\max}$, so the final configuration must have $n - 2$ chips at the center, 1 chip at the end of the single non-bidirected edge and 2 chips everywhere else. In this final configuration, every vertex has exactly one less chip than its outdegree.

Eriksson's proof of the claim, which we will not reproduce here, examines the chip-firing process from the initial configuration explicitly. The symmetry of the graph allows for the formulation of an inductive recursion. Asymptotically, the number of fires until stabilization is

$$\left(\frac{2n}{\sqrt{5}} + \tau\right)\tau^{(n-2)},$$

where τ is the golden ratio; see [Eri91].

6.5.4   Oriented spanning trees

Chapter 3 explored the many combinatorial connections to chip-firing starting with the connection to spanning trees. The appropriate notion of directed spanning tree which is equinumerous with critical configurations is a rooted arborescence; see Definition 6.5.1.

As in the undirected case, the enumeration result uses a Matrix-Tree Theorem; see e.g [Sta99].

Theorem 6.5.16 (Matrix-Tree Theorem for Directed Graphs). *For a directed graph G, the determinant of the reduced directed graph Laplacian (reduced at q) is equal to the number of arborescences of G rooted at q.*

Example 6.5.17. Consider the (non-Eulerian) directed graph of Example 6.5.7. In Example 6.5.7, the critical, z-superstables and G-parking functions are listed. There are precisely three of each. Figure 6.7 shows the corresponding arborescences rooted at q.

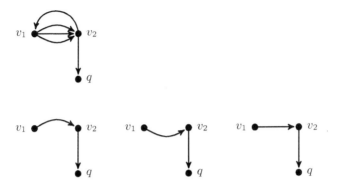

FIGURE 6.7: The three arborescences of the graph in Example 6.5.17.

The connection to the Laplacian leads to the following generalization of Theorem 3.1.3.

Theorem 6.5.18. *For a directed graph G with a sink q and at least one arborescence rooted at q, the following are equinumerous:*

- *Critical configurations*

- *z-superstable configurations*

- *G-parking functions*

- *Stable and allowed configurations*

- *Arborescences rooted at q*

Of course, one would like explicit bijections between arborescences and any of the collections listed above.

Both Merino's Theorem and the Cori–Le Borgne bijection from Chapter 3 have been extended to the *Eulerian* directed case. The critical polynomial incorporating an appropriate notion of level was formulated by Perrot and Pham [PP16]. Furthermore, Chan [Cha18] shows that this generating function is an evaluation of a generalization of the Tutte polynomial known as the greedoid Tutte polynomial.

6.6 Cartan matrices as M-matrices

Using the setup of Section 6.1, one can investigate chip-firing dynamics in new non-graphical contexts. Benkhart, Klivans and

Reiner [BKR16] studied the Cartan matrix of a finite root system as an avalanche-finite matrix. In order to state the main results, we start with a few preliminaries on the geometry and combinatorics of finite root systems; see e.g. [BB05] and [Dav08] for more on root systems.

For a vector $\alpha \in \mathbb{R}^n$, define the *reflection*

$$s_\alpha(x) = x - 2\langle \alpha^\vee, x \rangle \alpha,$$

for all $x \in \mathbb{R}^n$, where

$$\beta^\vee = \frac{2\beta}{\langle \beta, \beta \rangle}.$$

The *mirror of reflection* for s_α is the hyperplane with normal vector α. A finite root system corresponds to a collection of normal vectors to a closed system of mirrors:

Definition 6.6.1. A finite collection of vectors Φ in $\mathbb{R}^n$ is a *finite root system* if the following three conditions are satisfied:

1. Φ spans $\mathbb{R}^n$.

2. $s_\alpha(\Phi) = \Phi$ for all $\alpha \in \Phi$.

3. $\Phi \cap \mathbb{R}\alpha = \pm\alpha$ for all $\alpha \in \Phi$.

Furthermore a finite root system is called *crystallographic* if

$$\langle \alpha, \beta^\vee \rangle \in \mathbb{Z} \text{ for all } \alpha, \beta \in \Phi.$$

Let $\mathcal{H}$ be a *finite reflection arrangement* consisting of the hyperplanes perpendicular to a finite root system Φ. Fix any chamber F of $\mathcal{H}$, which we will refer to as the *fundamental chamber* and let Δ be the subcollection of roots that are normal to the bounding planes of F. The roots contained in Δ are called the *simple roots* of Φ (with respect to F); see Figure 6.8.

The crystallographic condition ensures that every element of Φ is an integer combination of the simple roots and that the $\mathbb{Z}$-span of Φ determines a well-defined lattice $Q(\Phi)$, the *root lattice* of Φ.

The *weight lattice* of a root system Φ is

$$P(\Phi) = \{v \in V \mid (v, \alpha^\vee) \in \mathbb{Z} \text{ for all } \alpha \text{ in } \Phi\}.$$

In the crystallographic case, the weight lattice $P(\Phi)$ contains the root lattice $Q(\Phi)$ as a sublattice. The *fundamental weights* span the extreme rays of the fundamental chamber; again see Figure 6.8.

The Cartan matrix records the coefficients of the expansion of the simple roots in terms of the fundamental weights.

Definition 6.6.2. For a finite crystallographic root system Φ, with choice of simple roots Δ, the *Cartan matrix* $C_\Delta = (c_{ij})$ is given by

$$c_{ij} = (\alpha_i, \alpha_j^\vee).$$

The *fundamental group* of Φ, is the cokernel $\operatorname{coker}(C^T)$ and can be reinterpreted as

$$\operatorname{coker}(C^T) \cong P(\Phi)/Q(\Phi).$$

The size of the cokernel $|\operatorname{coker}(C^T)|$ is called the *index of connection* for Φ. Our connection to chip-firing comes from the next proposition.

Proposition 6.6.3. *The Cartan matrix C of a finite, crystallographic, irreducible root system is an avalanche-finite matrix.*

The critical and superstable configurations defined by C will be stated in terms of dominant and minuscule weights. The *dominant* weights are the elements of $F \cap P(\Phi)$. And an element $\lambda \in P(\Phi)$ is *minuscule* if $\langle \lambda, \alpha^\vee \rangle \in \{-1, 0, 1\}$ for all $\alpha \in \Phi$.

Finally, the *Weyl vector* which is the sum of all the fundamental weights

$$\varrho = \sum_i^n \lambda_i,$$

will play the role of $\mathbf{c}_{\max}$. By identifying $\mathbb{Z}^n$ with $P(\Phi)$ we interpret the chip-firing process induced by C over the lattice $P(\Phi)$. For a choice of simple roots $\Delta = \{\alpha_1, \ldots, \alpha_n\}$ and an initial configuration $\mathbf{c}$, firing site i results in the configuration $\mathbf{c} - \alpha_i$.

Theorem 6.6.4 ([BKR16]). *Let C be the Cartan matrix of a finite, crystallographic, irreducible root system.*

1. *The superstable configurations are the zero vector $\mathbf{0}$ and the minuscule dominant weights λ.*

2. *The critical configurations are ϱ and $\varrho - \lambda$ for all minuscule dominant weights λ.*

3. *The burning configurations are the nonzero elements of the fundamental chamber that lie in the root lattice $Q(\Phi)$.*

Example 6.6.5. Consider the finite root system A_{n-1}, shown in Figure 6.8 for $n = 3$. Let

$$\alpha_i = \{e_{i+1} - e_i\}$$

be the simple roots. The fundamental chamber corresponding to this choice is

$$F = \{x \in \mathbb{R}^n \mid x_1 \leq \cdots \leq x_n\}.$$

The fundamental weights are labeled λ_i in the Figure. The Cartan matrix is

$$C = \begin{pmatrix} 2 & -1 \\ -1 & 2 \end{pmatrix}.$$

The entries of C give the expansion of the simple roots in terms of the fundamental weights:

$$\alpha_1 = 2\lambda_1 - \lambda_2 \quad \text{and} \quad \alpha_2 = -\lambda_1 + 2\lambda_2.$$

Consider chip-firing on a two-site system with the firing rule given by C. The critical configurations are $(0,1), (1,0), (1,1)$. The point $(1,1)$, for example, is identified with the Weyl vector $\varrho = \lambda_1 + \lambda_2$. The superstable configurations are $(1,0), (0,1), (0,0)$. The two non-zero configurations correspond to the minuscule dominant weights.

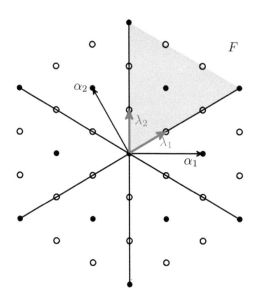

FIGURE 6.8: The reflection arrangement corresponding to the finite root system A_2. The simple roots are $\{\alpha_1, \alpha_2\}$ and the fundamental weights are $\{\lambda_1, \lambda_2\}$. The weight lattice consists of all the lattice points shown. The root lattice consists of the filled-in lattice points.

In the papers [BKR16] and [Gae16], the McKay–Cartan matrix of a faithful representation $\gamma : G \hookrightarrow GL_n(\mathcal{C})$ of a finite group G is similarly explored as an avalanche-finite matrix.

The papers [GHMP17a] and [GHMP17b] present a different connection between chip-firing and root systems as described in Chapter 5 Section 5.4. See [GHMP17a, Section 10] for the relationship between the two chip-firing processes.

6.7 M-pairings

It is sometimes the case that we would like to chip-fire over matrices that are not M-matrices. The combinatorial Laplacians of the next chapter will serve as a prime example. Guzman and Klivans [GK16] set up a general theory to chip-fire over any invertible integer matrix while maintaining much of the good behavior expected from a chip-firing process. The idea is to pair a given matrix with an M-matrix.

Let L be a non-singular $n \times n$ integer matrix. Let M be a non-singular $n \times n$ (real) M-matrix. We consider a chip-firing process for the pair (L, M) with n sites. The chip-firing process for the pair is constructed so that every initial configuration will eventually stabilize and so that local confluence and hence global confluence holds.

A chip configuration will no longer simply be a non-negative integer vector. Instead, define the following set:

Definition 6.7.1. For $N = LM^{-1}$, let

$$S^+ = \{Nx \mid Nx \in \mathbb{Z}^n,\ x \in \mathbb{R}^n_{\geq 0}\}.$$

The integer points inside S^+ are the *valid chip configurations* for the pair (L, M).

For a valid configuration $\mathbf{c} \in S^+$ firing a site i is defined to be the process which replaces the configuration $\mathbf{c}$ with $\mathbf{c} - L^T e_i$, namely subtracting the ith row of L from $\mathbf{c}$. A *legal* chip fire is one in which the resulting configuration stays within S^+:

$$\mathbf{c} - L^T e_i = \mathbf{c}' \in S^+.$$

A site i is *ready to fire* if firing site i results in a legal fire.

We think of S^+ as the cone of lattice points over which the chip-firing process operates. In the classical setting, S^+ is $\mathbb{Z}_{\geq 0}^n$, the integer points of the positive orthant. For a chip configuration on a graph, a legal firing is one in which the resulting configuration does not have negative entries, i.e. the resulting configuration remains within the positive orthant. Figures 6.9 and 6.10 demonstrate this geometry.

We have the following familiar definitions. A configuration is called *stable* if no site is ready to fire. A configuration is called *critical* if it is both stable and reachable from a sufficiently large initial configuration, where sufficiently large means that every site can legally fire.

A collection of sites can legally multi *cluster-fire* if $\mathbf{c} - L\mathbf{z}$ remains in S^+ where z_i records the number of times that site i fires. A configuration in which no non-empty set of sites can legally multi cluster-fire is called *superstable*.

The *sandpile group* of L is

$$\mathcal{S}(L) := \operatorname{coker}(L) = \mathbb{Z}^n / \operatorname{im}(L).$$

Two configurations $\mathbf{c}$ and $\mathbf{d}$ are *firing-equivalent* if $\mathbf{c} - \mathbf{d} = L\mathbf{z}$ for some $\mathbf{z}$.

The *energy* of a configuration $\mathbf{c} \in S^+$ is

$$||L^{-1}\mathbf{c}||_2^2.$$

An *energy minimizer* is a configuration in S^+ with the minimal energy among all firing equivalent configurations in S^+.

What is unfamiliar in this case is that stable, critical, superstable and energy minimizing configurations may easily have negative entries, as seen in Example 6.7.3.

Theorem 6.7.2 ([GK16]). *For an invertible integer matrix L and an M-matrix M, there exists unique critical, superstable and energy minimizing configurations per equivalence class of the $\operatorname{coker}(L)$ under the chip-firing process defined by the pair (L, M).*

Furthermore, the superstable and energy minimizing configurations coincide.

Unexpectedly, duality does not extend to this setting – superstable and critical configurations are not additive complements.

Figure 6.9 shows a representation of the non-negative orthant $\mathbb{Z}_{\geq 0}^n$. Consider the chip-firing process on a finite graph with a sink. Generically, an initial configuration is an integer point with many possible legal firings. Legal firings are performed until a configuration inside the rectilinear box defined by $\mathbf{c}_{max}$ is reached. The process then stops before firing outside of the positive orthant. As in Figure 2.9, the critical configurations can be seen as an upward closed collection of stable configurations clustered at $\mathbf{c}_{max}$. The superstable configurations can be seen as a downward closed collection clustered at the origin.

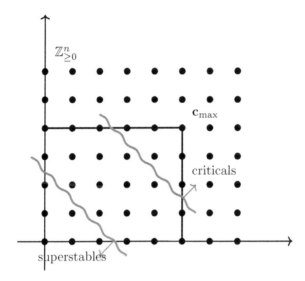

FIGURE 6.9: The geometry of valid configurations in the graphical or more general M-matrix case.

Figure 6.10 shows a representation of the cone S^+. A legal fire is defined as one which does not result in a configuration outside of S^+. The chip-firing process starts with an initial configuration with many possible legal fires and stops before any further firings will result in leaving S^+. While there is a well-defined collection of stable configurations, the relation between and geometry of the critical and superstable configurations is not well understood.

Example 6.7.3 ([GK16, Example 2.5]). Let

$$L = \begin{pmatrix} 2 & -1 & 1 \\ -1 & 2 & -1 \\ 1 & -1 & 2 \end{pmatrix} \text{ and } M = \begin{pmatrix} 3 & -1 & -1 \\ -1 & 3 & -1 \\ -1 & -1 & 3 \end{pmatrix}.$$

Then $N = \begin{pmatrix} 1 & .25 & .75 \\ -.25 & .5 & -.25 \\ .75 & .25 & 1 \end{pmatrix}$. The valid configurations, the integer points of S^+, consist of integer vectors of the form Nx such that $x \geq 0$.

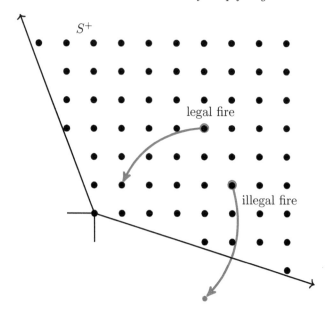

FIGURE 6.10: The geometry of chip-firing over S^+ for a matrix pairing (L, M).

Elements of S^+ include, e.g.

$$(0, 0, 0),$$
$$(1, 0, 1),$$
$$(3, -1, 4).$$

On the other hand,

$$(0, 0, 1) \text{ and } (1, -1, 1)$$

are not in S^+.

The critical configurations are:

$$(4, -1, 4),$$
$$(4, 0, 4),$$
$$(5, 0, 5),$$
$$(5, -1, 5).$$

The superstable configurations are:

$$(1, 0, 1),$$
$$(1, 1, 1),$$
$$(0, 0, 0),$$
$$(2, 1, 2).$$

Given a matrix L, the matrix M can be any M-matrix. Therefore, there are many chip-firing processes that one can define for a fixed L. For any pairing, the cone S^+ may change, but the firing rule remains the same and is defined only in terms of L. The change of S^+ does change the collections of critical and superstable configurations. Two interesting special cases arise at the extremes of these choices.

First, suppose that L is an M-matrix, e.g. a reduced graph Laplacian. Then L can be paired with itself. In the chip-firing process (L, L),

$$N = LL^{-1} = I, \ S^+ = \mathbb{Z}_{\geq 0}^n,$$

and we recover usual graphical chip-firing as a special case.

For the other extreme, consider pairing a matrix L with the identity matrix I, which is an M-matrix. In fact, I is the reduced graph Laplacian for the star graph with the star vertex as the sink. In this case, the critical and superstable configurations coincide and are precisely the integer points of the fundamental parallelepiped of the lattice generated by the columns of L; see [GK16].

6.8 Exercises

Exercise 6.8.1. *Prove the equivalences of Theorem 6.1.3.*

Exercise 6.8.2. *Prove Proposition 6.4.5, for a given M-matrix there is a unique minimal burning configuration.*

Exercise 6.8.3. *Let*

$$M = \begin{pmatrix} 2 & -2 & -5 \\ 0 & 3 & -1 \\ 0 & 0 & 12 \end{pmatrix}.$$

   1. *Confirm that M is an M-matrix.*

   2. *Find the critical and superstable configuration associated to M.*

Exercise 6.8.4. *(Open)*

   1. *Is the space of M-matrices convex?*

   2. *Given a matrix L, what is the closest M-matrix?*

Exercise 6.8.5. *Prove the equivalences of Theorem 6.3.6.*

Exercise 6.8.6. *Prove that for directed graphs, the sandpile group is invariant to the choice of sink if G is Eulerian.*

Exercise 6.8.7. *Prove Proposition 6.5.5, that χ-superstables, z-superstables, and G-parking functions are the same for Eulerian graphs G.*

Exercise 6.8.8. *Fill in the details to prove Eriksson's bound, Theorem 6.5.15.*

Exercise 6.8.9. *Prove that every critical configuration is allowed.*

Exercise 6.8.10. *Show that for general non-Eulerian directed graphs, maximal superstable configurations may not be maximum.*

Exercise 6.8.11. *Prove Proposition 6.5.13 that in a strongly connected graph, if each site fires at least h_v times from some initial configuration $\mathbf{c}$, then the process will never stabilize.*

Exercise 6.8.12. *Let Φ be the finite root system B_3 consisting of hyperplanes $\{x_i = x_j, x_i = 0\} 1 \le i, j \le 3\}$ and simple roots:*

$$\{e_1 - e_2, e_2 - e_3, e_3\}.$$

Find the critical and superstable configurations for the Cartan matrix of Φ.

Exercise 6.8.13. *Let L be non-singular integer matrix. Let M be an M-matrix and D a non-negative diagonal matrix. Prove that the critical and superstable configurations of the pair (L, M) and the pair (L, DM) are the same.*

Exercise 6.8.14. *Prove that for the pair (L, I), the critical configurations are the same as the superstable configurations and equal to the integers points inside the fundamental parallelepiped of the lattice generated by the columns of L.*

Exercise 6.8.15. *Let L be the matrix of Exercise 6.8.3. Find the critical and superstable configurations for the pair (L, I).*

Chapter 7

Higher Dimensions

In the previous chapter, we considered chip-firing processes on abstract networks with highly structured M-matrix dynamics. Here, we focus on a specific class of networks which come equipped with a natural operator but the operator is not an M-matrix.

Specifically, in this chapter, we chip-fire on topological cell complexes using dynamics defined by the combinatorial Laplacian. This generalizes the chip-firing process from graphs to higher dimensions.

In the higher-dimensional chip-firing model:

- Instead of chips firing from vertices to vertices along edges, flow is diverted from cells of codimension-one to cells of codimension-one along cells of top dimension.

- Firing classes are enumerated by cellular spanning trees.

- The sandpile group is the finite abelian group resulting from taking the torsion part of the cokernel of a reduced combinatorial Laplacian.

- The sandpile and newly defined co-sandpile groups are isomorphic to the discriminant groups of the cut and flow lattices of the cell complex.

7.1   Illustrative examples

The chip-firing process on graphs starts with a value at each vertex and the chip-firing rule disperses chips in a way that alters the values of a vertex and its neighbors.

In the higher-dimensional case, the chip-firing process starts with a

value at each cell of a fixed dimension. Firing a cell alters the value of
a cell and its neighbors. But, the chip-firing rule is now more naturally
described as a diversion of flow.

Consider the 2-simplex pictured in Figure 7.1. (Background on simplicial and cellular complexes will be presented in Section 7.2.) The complex consists of one 2-dimensional cell (the triangle), three 1-dimensional
cells (the edges) and three 0-dimensional cells (the vertices).

Two-dimensional chip-firing starts with values at each edge. These
values are represented by directed arrows on the edges and can be
thought of as an amount of flow traveling along the edge. For example, in Figure 7.1 three units of flow are being sent from vertex v_1 to
vertex v_3.

Firing an edge e results in re-routing flow across all 2-cells which contain e to neighboring edges. In our example, the edge with three units
of flow is contained in a single 2-cell. Firing this edge diverts one unit
of flow across the cell and along the other two edges of the cell boundary.

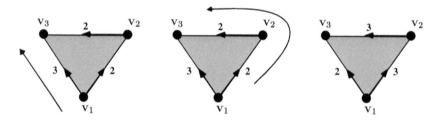

FIGURE 7.1: In two dimensions, flow is diverted from one edge to
neighboring edges.

After firing the edge $v_1 v_3$ the same amount of flow is being sent from
vertex v_1 to vertex v_3 but along a different path.

If an edge in a complex is contained in more than one 2-cell, then
firing the edge results in diverting one unit of flow across each 2-cell; see
Figure 7.2.

For another example, consider the two-dimensional cubical grid. A
small patch is shown in Figure 7.3. This complex is the two-dimensional
analogue of the graph of the one-dimensional lattice (an infinite path).
Again, chip-firing diverts flow along edges to neighboring edges.

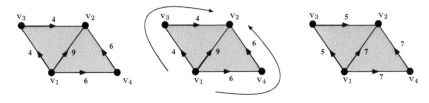

FIGURE 7.2: Firing the edge $v_1 v_2$ diverts flow across two 2-cells.

Suppose our initial configuration consists of three units of flow on four edges in a counter clockwise orientation and zero flow elsewhere, as seen in the leftmost grid of Figure 7.3. Each edge is contained in exactly two 2-cells. Firing any of the edges with positive flow results in diverting flow across the two 2-cells that contain the edge.

We interpret flow in opposite directions as canceling. For example, firing the edge oriented from right to left in the initial configuration yields the topmost configuration of the second column. One unit of flow is re-routed upwards to edges where there had previously not been any flow. One unit is re-routed downwards around the locations of the initial non-zero configuration but in the opposite orientation.

In this example, since each edge is contained in exactly two squares, we fire edges until all edges have at most one unit of flow. All possible resulting configurations and intermediate steps are shown in Figure 7.3. With this choice of firing rule, the process terminates (stabilizes) but clearly not in a unique configuration.

Already in these two examples, we can see interesting differences between higher-dimensional chip-firing and graphical chip-firing:

- If we consider the total amount of flow as the sum of all values on all edges, then flow is no longer conserved. In Figure 7.1, the total amount of flow in the initial configuration is 7 but the total amount of flow in the resulting configuration is 8. Instead of total flow, two-dimensional chip-firing conserves the quantity

$$\text{inflow}(v) - \text{outflow}(v)$$

  at each vertex v. This is consistent with the idea that flow is simply being rerouted.

- Flows are oriented. Oppositely oriented flows can cancel. A negative flow can be interpreted as a positive flow in the other direction. This makes stabilization problematic – when should the process terminate?

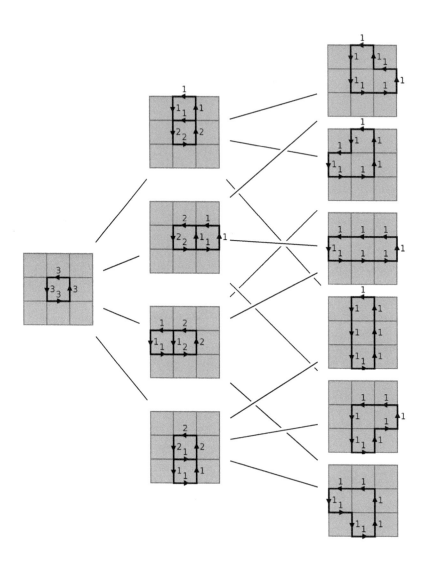

FIGURE 7.3: Two-dimensional chip-firing from an initial configuration consisting of a circulation of 3 units of flow.

- Since firing one edge can decrease the amount of flow on neighboring edges, local confluence also presents a difficulty. We will see that the operators defining the firing rule are not M-matrices.

- Our examples thus far have not had a sink. The sink in dimension two will not be a fixed single edge, instead the sink will be a certain collection of edges.

By using the M-pairing construction from the previous chapter, we will recover much of the good behavior expected from chip-firing for cell complexes with a sink.

7.2   Cell complexes

In this section and the next, we review and gather the material from topological combinatorics that we need in order to formalize higher-dimensional chip-firing. Good references for additional background on these topics include [Hat02], [Sta96a], and [Gib10].

Definition 7.2.1. An abstract *simplicial complex* Δ on ground set V is a non-empty collection of subsets of V such that

$$\text{if } \sigma \subseteq \tau \text{ and } \tau \in \Delta, \text{ then } \sigma \in \Delta.$$

The elements of Δ are the *cells* or *faces* of Δ.

The *dimension* of a face σ is $\dim(\sigma) = |\sigma| - 1$.

The *k-dimensional skeleton* or *k-skeleton* $\Delta_{(k)}$ is the subcomplex consisting of all faces of dimension $\leq k$.

A *facet* is a face of maximal dimension. A complex is *pure* if all facets have the same dimension. A *ridge* is a face of codimension one.

The *f-vector* $f(\Delta)$ records the number of faces of each dimension $f(\Delta) = (f_{-1}(\Delta), f_0(\Delta), \ldots, f_d(\Delta))$, where $f_k(\Delta)$ is the number of faces of dimension k.

Note that, by definition, the empty set is considered a face of all simplicial complexes.

Example 7.2.2. A finite simple graph is a one-dimensional simplicial complex with ground set equal to the vertex set. The edges are the one-dimensional faces, the vertices are the zero-dimensional faces, and the empty set is regarded as a face of dimension -1.

Simplicial complexes are special cases of hypergraphs: if we consider the faces of a simplicial complex as hyperedges, then a simplicial complex is a hypergraph whose collection of hyperedges is closed under taking subsets.

Every simplicial complex has a *geometric realization*. The faces of dimension d are realized as convex hulls of $d + 1$ affinely independent points. A 0-dimensional face is a point, a 1-dimensional face is a line segment, a 2-dimensional face is a triangle, etc. An abstract simplicial complex is realized as a geometric space by gluing together simplices along common smaller-dimensional simplices. A d-dimensional simplicial complex may always be realized in $\mathbb{R}^{2d+1}$, i.e. it is embeddable in no more than $2d + 1$ dimensions.

More generally, a *CW-complex* (or simply *cell complex*) consists of cells homeomorphic to topological disks that are not necessarily simplices. Examples include a polygon with $n > 3$ sides or the two-dimensional polyhedral complex which is the boundary of a dodecahedron. We will primarily be interested in simplicial complexes, but the cubical grid of Section 7.1 is a non-simplicial example.

A cell complex can be specified geometrically or it can be specified combinatorially in terms of its face poset.

Definition 7.2.3. The *face poset* of a cell complex Σ is the partial order on faces of Σ where a face σ is covered by a face τ if $\sigma \subset \tau$ and no other face sits between them: there does not exist $\gamma \in \Sigma$ such that $\sigma \subset \gamma \subset \tau$.

Example 7.2.4. Let Δ be the boundary complex of the tetrahedron. Then Δ is a two-dimensional complex consisting of four 2-cells, six 1-cells, and four 0-cells. Figure 7.4 shows a geometric realization of Δ and the face poset of Δ. Each face has been identified with its collection of vertices.

Example 7.2.5. Let Σ be the cubical complex shown in Figure 7.4. Then Σ is a two-dimensional non-simplicial complex consisting of three 2-cells, nine 1-cells, and seven 0-cells. In this example, vertices have been identified with binary string coordinates. The star stands for all real values between 0 and 1. Suppose the complex were realized geometrically in $\mathbb{R}^3$ with the vertices at coordinates equal to their labels. Then, for example, the edge between 010 and 110 would be all points in $\mathbb{R}^3$ of the form $(*, 1, 0)$.

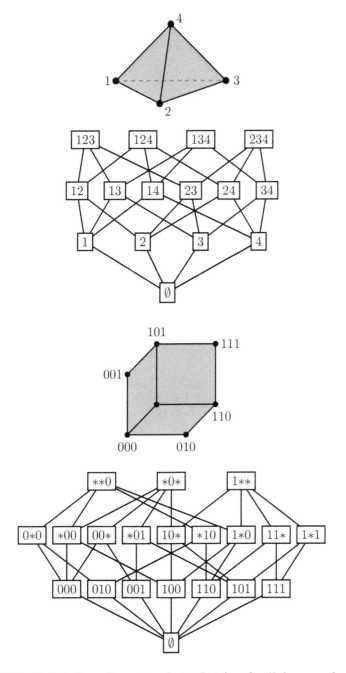

FIGURE 7.4: Two-dimensional simplicial and cellular complexes.

7.3   Combinatorial Laplacians

Let Δ be a d-dimensional simplicial complex. As with the oriented incidence matrix of a graph, there are maps relating faces of adjacent dimensions. Let the vertices of Δ be ordered $v_0, v_1, \ldots, v_n$.

The kth *oriented incidence matrix* of Δ is a $f_{k-1} \times f_k$ matrix ∂_k defined as follows:

$$(\partial_k)_{\sigma\tau} = \begin{cases} (-1)^j & \text{if } \tau = \{v_0, v_1, \ldots, v_k\}, \sigma = \tau \backslash \{v_j\} \\ 0 & \text{otherwise.} \end{cases} \tag{7.1}$$

We think of the ordered vertices as inducing an order on all faces of Δ. The sign of ∂_k corresponding to faces σ and τ reflects their relative orientations. A sign of $+1$ reflects that σ and τ are consistently oriented. A sign of -1 reflects that σ and τ are oppositely oriented.

An oriented simplicial complex is shown in Example 7.4.1. The orientation of a two-dimensional face is denoted by a clockwise or counterclockwise arrow. The edge $v_3 v_5$, for example, is consistently oriented with the triangle $v_1 v_3 v_5$, whereas the edge $v_1 v_5$ is not consistently oriented with the triangle $v_1 v_3 v_5$.

The *simplicial chain group* of Σ, $C_k(\Sigma; \mathbb{Z})$, consists of all formal linear combinations of k-dimensional faces known as chains. The kth oriented incidence matrix extends linearly to the chain group and is known as the kth *boundary map*:

$$\partial_k : C_k(\Delta; \mathbb{Z}) \to C_{k-1}(\Delta; \mathbb{Z}).$$

The composition of two adjacent boundary maps ∂_k and ∂_{k-1} is equal to zero, hence the sequence of chain groups and boundary maps forms an algebraic chain complex.

Example 7.3.1. Let Δ be the boundary of the tetrahedron of Example 7.2.4. The chain complex of Δ consists of three boundary maps:

$$C_2 \xrightarrow{\partial_2} C_1 \xrightarrow{\partial_1} C_0 \xrightarrow{\partial_0} \mathbb{Z}.$$

The 2nd boundary map relates the four triangular faces to the six edges.

$$\partial_2 = \begin{array}{c} \\ 12 \\ 13 \\ 14 \\ 23 \\ 24 \\ 34 \end{array} \begin{array}{cccc} 123 & 124 & 134 & 234 \\ \left(\begin{array}{cccc} 1 & 1 & 0 & 0 \\ -1 & 0 & 1 & 0 \\ 0 & -1 & -1 & 0 \\ 1 & 0 & 0 & 1 \\ 0 & 1 & 0 & -1 \\ 0 & 0 & 1 & 1 \end{array} \right) \end{array}$$

$$C_2(\Delta; \mathbb{Z}) \longrightarrow C_1(\Delta; \mathbb{Z}).$$

The 1st boundary map relates edges to vertices; it is the oriented incidence matrix of the graph formed by the edges of Δ, which in this case is K_4.

$$\partial_1 = \begin{array}{c} \\ 1 \\ 2 \\ 3 \\ 4 \end{array} \begin{array}{cccccc} 12 & 13 & 14 & 23 & 24 & 34 \\ \left(\begin{array}{cccccc} 1 & 1 & 1 & 0 & 0 & 0 \\ -1 & 0 & 0 & 1 & 1 & 0 \\ 0 & -1 & 0 & -1 & 0 & 1 \\ 0 & 0 & -1 & 0 & -1 & -1 \end{array} \right) \end{array}$$

$$C_1(\Delta; \mathbb{Z}) \longrightarrow C_0(\Delta; \mathbb{Z}).$$

The 0th boundary map is given by the all 1s vector. The map ∂_0 appeared in Chapter 4 in the construction of the sandpile group; see Definition 4.2.1.

$$C_0(\Delta; \mathbb{Z}) \quad \overset{\partial_0 = \emptyset \begin{array}{cccc} 1 & 2 & 3 & 4 \\ \left(\begin{array}{cccc} 1 & 1 & 1 & 1 \end{array} \right) \end{array}}{\longrightarrow} \mathbb{Z}.$$

Definition 7.3.2. For a simplicial complex Δ, the *kth homology group* of Δ is

$$H_k(\Delta; \mathbb{Z}) = \ker \partial_k / \operatorname{im} \partial_{k+1}.$$

Homology groups are finitely generated abelian groups hence they can be written as:

$$H_k(\Delta; \mathbb{Z}) \cong \mathbb{Z}^{\beta_k} \oplus (\mathbb{Z}/d_1\mathbb{Z}) \oplus (\mathbb{Z}/d_2\mathbb{Z}) \oplus \cdots \oplus (\mathbb{Z}/d_m\mathbb{Z}).$$

The integer β_k, the number of copies of $\mathbb{Z}$ in the free part of the homology group, is the *kth Betti number* of Δ,

$$\beta_k(\Delta) = \dim_{\mathbb{Q}} H_k.$$

The finite summands of the homology group are called the *torsion* part of the homology.

The *simplicial cochain group* of Σ, $C_n^* = \mathrm{Hom}(C_n; \mathbb{Z})$ is the dual space of homomorphisms from C_n to $\mathbb{Z}$. Having identified the cochains with chains, the *coboundary map* is given by the transpose of the boundary map:

$$\partial_k^T \; : \; C_{k-1}(\Delta; \mathbb{Z}) \to C_k(\Delta; \mathbb{Z}).$$

Finally, we come to the operator that will dictate chip-firing moves in higher dimensions.

Definition 7.3.3. For a simplicial complex Δ, the *k-dimensional combinatorial Laplacian of Δ* is

$$L_k = \partial_k \partial_k^T \; : \; C_{k-1}(\Delta; \mathbb{Z}) \to C_{k-1}(\Delta; \mathbb{Z}).$$

Example 7.3.4. Again let Δ be the boundary complex of the tetrahedron. The one- and two-dimensional Laplacians of Δ are:

$$L_2 = \partial_2 \partial_2^T \; : \; C_1(\Delta; \mathbb{Z}) \to C_1(\Delta; \mathbb{Z})$$

$$=
\begin{array}{c}
\begin{array}{cccccc}
12 & 13 & 14 & 23 & 24 & 34
\end{array} \\
\begin{array}{c}
12 \\ 13 \\ 14 \\ 23 \\ 24 \\ 34
\end{array}
\left(
\begin{array}{cccccc}
2 & -1 & -1 & 1 & 1 & 0 \\
-1 & 2 & -1 & -1 & 0 & 1 \\
-1 & -1 & 2 & 0 & -1 & -1 \\
1 & -1 & 0 & 2 & -1 & 1 \\
1 & 0 & -1 & -1 & 2 & -1 \\
0 & 1 & -1 & 1 & -1 & 2
\end{array}
\right).
\end{array}
$$

$$L_1 = \partial_1 \partial_1^T \; : \; C_0(\Delta; \mathbb{Z}) \to C_0(\Delta; \mathbb{Z})$$

$$=
\begin{array}{c}
\begin{array}{cccc}
1 & 2 & 3 & 4
\end{array} \\
\begin{array}{c}
1 \\ 2 \\ 3 \\ 4
\end{array}
\left(
\begin{array}{cccc}
3 & -1 & -1 & -1 \\
-1 & 3 & -1 & -1 \\
-1 & -1 & 3 & -1 \\
-1 & -1 & -1 & 3
\end{array}
\right).
\end{array}
$$

In general, for the combinatorial Laplacian:

- An off-diagonal entry corresponding to faces σ_1 and σ_2 is equal to 1 if σ_1 and σ_2 are contained in a common face τ and the relative orientation of σ_1 to τ is the same as the relative orientation of σ_2 to τ.

- An off-diagonal entry corresponding to faces σ_1 and σ_2 is equal to -1 if σ_1 and σ_2 are contained in a common face τ and the relative orientation of σ_1 to τ is opposite that of the relative orientation of σ_2 to τ.

- An off-diagonal entry corresponding to faces σ_1 and σ_2 is equal to 0 if σ_1 and σ_2 are not contained in a common face.

- The diagonal entries are the degrees of faces. For the tetrahedron, L_2 has all 2s along the diagonal. This reflects the fact that every edge is contained in exactly 2 facets.

Combinatorial Laplacian operators (also known as Hodge Laplacians) seem to have first appeared in the work of Eckmann [Eck44] on finite-dimensional Hodge theory. As the name suggests, they are discrete versions of the Laplacian operators on differential forms on a Riemannian manifold. In fact, Dodziuk and Patodi [DP96] showed that for suitably nice triangulations of a manifold, the eigenvalues of the discrete Laplacian of the triangulation converge (in an appropriate sense) to those of the analytic Laplacian.

The Laplacian as defined above, $\partial_k \partial_k^T$, is often referred to as the *up-down* Laplacian L_k^{UD}. One can similarly define the *down-up* Laplacian $L_k^{DU} = \partial_{k-1}^T \partial_{k-1}$ and the *total Laplacian*:

$$L_k = \partial_k \partial_k^T + \partial_{k-1}^T \partial_{k-1}.$$

See, for example, [DR02] for further discussion on the various combinatorial Laplacians.

We have defined the combinatorial Laplacian only for simplicial complexes but the Laplacian can be constructed for any cell complex as

$$L = \partial \partial^T.$$

The difficulty is that there is no simple formula for ∂ analogous to Equation 7.1 for arbitrary cell complexes.

7.4   Chip-firing in higher dimensions

Chip-firing in higher dimensions was introduced by Duval, Klivans and Martin in [DKM13]. As in the graphical case, we use the Laplacian to define a discrete diffusion.

Let Δ be a finite oriented d-dimensional cell complex with d-dimensional Laplacian L. Let $m = f_{d-1}(\Delta)$ be the number of ridges of Δ.

- A *chip configuration* for Δ is any integer vector

$$\mathbf{c} = (c_1, c_2, \ldots, c_m) \in \mathbb{Z}^m.$$

  We interpret $\mathbf{c}$ as recording the amount of flow along each ridge of the complex. A positive value at c_i represents a flow of magnitude $|c_i|$ in the same orientation as the orientation of the ith face. A negative value at c_i represents a flow of magnitude $|c_i|$ in the opposite orientation as the orientation of the ith face.

- From a configuration $\mathbf{c}$, *firing* the ith face results in the configuration

$$\mathbf{c}' = \mathbf{c} - L e_i.$$

  The process diverts flow from the ith face to neighboring faces. Flow is decreased at the ith face. Flow may increase or decrease at neighboring faces.

Example 7.4.1. Let Δ be the two-dimensional simplicial complex consisting of three triangles with the standard orientation as shown in Figure 7.5. The boundary maps and two-dimensional Laplacian are given below. The ordering of the faces is lexicographic.

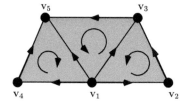

FIGURE 7.5: A two-dimensional complex with the standard orientation.

$$C_2 \xrightarrow{\begin{pmatrix} 1 & 0 & 0 \\ -1 & 1 & 0 \\ 0 & 0 & 1 \\ 0 & -1 & -1 \\ 1 & 0 & 0 \\ 0 & 1 & 0 \\ 0 & 0 & 1 \end{pmatrix}} C_1 \xrightarrow{\begin{pmatrix} 1 & 1 & 1 & 1 & 0 & 0 & 0 \\ -1 & 0 & 0 & 0 & 1 & 0 & 0 \\ 0 & -1 & 0 & 0 & -1 & 1 & 0 \\ 0 & 0 & -1 & 0 & 0 & 0 & 1 \\ 0 & 0 & 0 & -1 & 0 & -1 & -1 \end{pmatrix}} C_0$$

$$L_2 = \begin{pmatrix} 1 & -1 & 0 & 0 & 1 & 0 & 0 \\ -1 & 2 & 0 & -1 & -1 & 1 & 0 \\ 0 & 0 & 1 & -1 & 0 & 0 & 1 \\ 0 & -1 & -1 & 2 & 0 & -1 & -1 \\ 1 & -1 & 0 & 0 & 1 & 0 & 0 \\ 0 & 1 & 0 & -1 & 0 & 1 & 0 \\ 0 & 0 & 1 & -1 & 0 & 0 & 1 \end{pmatrix}.$$

Consider the initial configuration:

$$(2, 2, 2, 2, 1, 5, 4)$$

as shown in the first complex of Figure 7.6. Firing the edge $v_1 v_2$ results in the configuration on the right. One unit of flow is diverted along the edges $v_1 v_3$ and $v_3 v_2$. This has the effect of decreasing the magnitude of flow on the edge $v_3 v_2$. Note that the quantity inflow(v) − outflow(v) is conserved at each vertex v. Figure 7.6 shows two additional fires, of edge $v_1 v_5$ and edge $v_1 v_2$.

Figure 7.7 shows the last flow configuration of Figure 7.6 represented in two different ways. In the first, an edge has a negative flow. In the second, the orientation of the edge and sign of the flow value have been swapped. The ability to interpret a negative flow as a positive flow in the opposite direction makes it difficult to define a stopping condition.

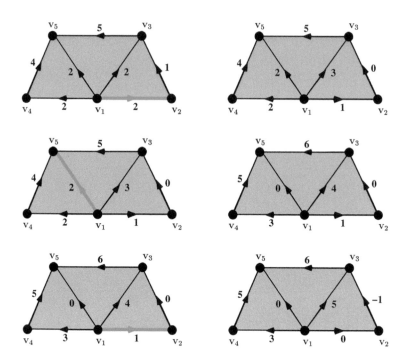

FIGURE 7.6: Flow-firing on a two-dimensional complex.

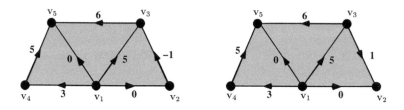

FIGURE 7.7: Two equivalent flow configurations.

7.5 The sandpile group

Let Δ be a finite d-dimensional complex with d-dimensional Laplacian L. Two flow configurations $\mathbf{c}$ and $\mathbf{d}$ on $(d-1)$-dimensional faces of Δ are *firing equivalent* if

$$\mathbf{c} - \mathbf{d} = L^T \mathbf{z}$$

for some integer vector $\mathbf{z}$.

Definition 7.5.1. For a d-dimensional complex Δ, with d-dimensional Laplacian, the d-dimensional *sandpile group* of Δ is

$$\mathcal{S}(\Delta) = \ker \partial_{d-1} / \operatorname{im} L_d.$$

The definition of the sandpile group is in terms of the kernel of the boundary map. In the graphical case, chip configurations in the kernel of ∂_0 are configurations with the total number of chips equal to 0. In higher dimensions, we call flow configurations in $\ker(\partial_{d-1})$ *conservative flows*.

In our two-dimensional examples, conservative flows are those with

$$\operatorname{inflow}(v) - \operatorname{outflow}(v) = 0$$

at each vertex.

The sandpile group is thus the group of equivalence classes of conservative flows under the Laplacian firing equivalence.

Without a stabilization operator, we cannot define the sandpile group as an additive group of configurations as in Chapter 4. In order to further investigate the sandpile group, we first define higher-dimensional spanning trees.

7.6 Higher-dimensional trees

Thus far our chip-firing model has been akin to chip-firing on a graph without a sink. For a d-dimensional complex Δ, the sink will not simply be a fixed face of dimension $d-1$. Instead, the sink will be a certain collection of faces. These collections are the higher-dimensional trees of Δ.

Higher-dimensional trees have their origins in the work of Bolker [Bol76] and Kalai [Kal83]. The theory has been further advanced more recently and we follow [DKM16] for our introduction.

We define higher-dimensional trees in analogy to graphical spanning trees. Consider the following extended definition of graphical trees:

Definition 7.6.1. A graph G on n vertices with m edges is a *tree* if it satisfies any of the following equivalent properties:

1. G is connected and acyclic.

2. G is connected and $m = n - 1$.

3. G is acyclic and $m = n - 1$.

4. Every pair of vertices in G is connected by exactly one path.

5. G is a maximally acyclic graph.

6. G is a minimally connected graph.

7. For any (hence every) orientation of G, the columns of the incidence matrix ∂_G are a basis for the space

$$\mathbb{R}_0^n = \{v \in \mathbb{R}^n \mid v_1 + \cdots + v_n = 0\}.$$

A graph G is a *forest* if it satisfies any of the following equivalent properties:

1. G is acyclic.

2. Every pair of vertices in G is connected by at most one path.

3. The columns of the incidence matrix ∂_G of G are linearly independent.

4. Every connected component of G is a tree.

Definition 7.6.2. A *spanning tree* (forest) of a graph G is a tree (forest) T such that $T \subset G$ and T contains all vertices of G.

Next we define higher-dimensional trees and forests. Most of the conditions are stated in homological terms, but the last of each list is stated in terms of linear algebra. This will allow us to define higher-dimensional trees in terms of matroid theory.

Definition 7.6.3 ([DKM16]). Let Δ be a cell complex of dimension d with $\beta_{d-1}(\Delta) = 0$[1] and let $\Upsilon \subset \Delta$ contain the skeleton Δ_{d-1}. Then Υ is a *spanning tree* of Δ if it satisfies any of the following equivalent properties:

1. $\beta_{d-1}(\Upsilon) = \beta_{d-1}(\Delta)$ and $\beta_d(\Upsilon) = 0$.

2. $\beta_{d-1}(\Upsilon) = \beta_{d-1}(\Delta)$ and $|\Upsilon_d| = |\Delta_d| - \beta_d(\Delta)$.

3. $\beta_d(\Upsilon) = 0$ and $|\Upsilon_d| = |\Delta_d| - \beta_d(\Delta)$.

4. Every element of $\ker \partial_{d-1}(\Delta)$ is the boundary of exactly one d-chain in Υ.

5. Υ is maximal among the spanning subcomplexes of Δ with $\beta_d(\Upsilon) = 0$.

6. Υ is minimal among the spanning subcomplexes of Δ with $\beta_{d-1}(\Upsilon) = \beta_{d-1}(\Delta)$.

7. The columns of $\partial_d(\Upsilon)$ are a vector space basis for the colspace $\partial_d(\Delta)$.

The complex Υ is a *spanning forest* if it satisfies any of the following equivalent properties:

1. $\beta_d(\Upsilon) = 0$.

2. Every element of $\ker \partial_{d-1}(\Delta)$ is the boundary of at most one d-chain in Υ.

3. The columns of $\partial_d(\Upsilon)$ are linearly independent.

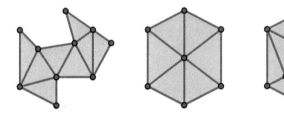

FIGURE 7.8: Examples of trees and forests.

Example 7.6.4. The first two complexes of Figure 7.8 are two-dimensional trees. Note that in the second complex the dual graph of

---

[1]The equivalences of Definition 7.6.3 continue to hold with only slight modification in the more general case that $\beta_{d-1} \neq 0$; see [DKM16] for details.

simplicial cells does not form a graphical tree as is often required in definitions of hypergraph trees. The third complex is a two-dimensional forest. As seen here, in dimensions greater than one, forests do not necessarily consist of disjoint unions of trees.

Example 7.6.5. Consider our running example of the boundary complex of a tetrahedron Δ. There are four spanning trees of Δ. Any collection of three 2-cells forms a spanning tree. We can think of these trees as being formed by removing any one boundary cell, which pierces the two-dimensional homology cycle of Δ.

This example extends to the boundary complex of any polytope: any collection of all but one boundary cell forms a spanning tree of the complex.

Boundary complexes of polytopes are the higher-dimensional analogues of cycle graphs. The removal of any one edge from a cycle graph gives a spanning tree.

Condition 7 of the definition of trees makes it clear that higher-dimensional spanning trees are the bases of a realizable matroid. For a cell complex Σ there is a natural matroid formed by the columns of the boundary map ∂ of Σ.

Definition 7.6.6. Let Σ be a d-dimensional complex with top boundary map ∂. The *cellular matroid* of Σ, $M(\Sigma)$, is the matroid with ground set equal to the collection of facets of Σ. A subset of facets is a basis of $M(\Sigma)$ if the corresponding subset of columns of ∂ is a column basis of ∂.

Simplicial matroids, in particular for the complete complex $\Delta_{(n,k)}$ were investigated by Cordovil and Lindström; see [CL87].

For a d-dimensional cell complex Σ with $\beta_{d-1}(\Sigma) = 0$, a subcomplex $\Upsilon \subset \Sigma$ which contains the skeleton $\Sigma_{(d-1)}$ is a cellular spanning tree of Σ if the collection of facets of Υ is a basis for the simplicial matroid $M(\Sigma)$.

Example 7.6.7. Let Δ be the boundary of the tetrahedron. The matrix $\partial_2(\Delta)$ is shown in Example 7.3.1. The first three columns form a column basis. Therefore $\{123, 124, 134\}$ is a basis of the matroid $M(\Delta)$. In fact any three columns of ∂_2 form a column basis. $M(\Delta)$ is the uniform matroid $U_{4,3}$ with four bases, which correspond to the four simplicial spanning trees of Δ.

7.6.1 Enumeration of trees

The number of spanning trees of a complex turns out not to be a nice quantity. Higher-dimensional trees are more naturally enumerated with a torsion weighting as we will see in the results of this section.

Definition 7.6.8. For a d-dimensional complex Δ, the *torsion-weighted tree number*, τ_d, is given as follows

$$\tau_d(\Delta) = \sum_{\Upsilon \in \mathcal{T}(\Delta)} |H_{d-1}(\Upsilon)|^2.$$

Torsion-free trees contribute 1 to the sum, but trees with torsion contribute with potentially large multiplicity.

In order to state a higher-dimensional Matrix-Tree Theorem, we also need the following quantity. For a d-dimensional complex Σ with d-Laplacian L, let

$$\pi_d = \lambda_1 \lambda_2 \cdots \lambda_m$$

be the product of the non-zero eigenvalues of L.

Theorem 7.6.9 (The Cellular Matrix-Tree Theorem [DKM09]).
Let Δ be a d-dimensional complex such that $H_k(\Delta; \mathbb{Z}) = 0$ for all $k < d$. Let Υ be a spanning tree of $\Delta_{(d-1)}$, then

$$\tau_d = \det L_\Upsilon,$$

where L_Υ is the Laplacian restricted to faces not in Υ.
Alternatively,

$$\tau_d = \frac{\pi_d}{\tau_{d-1}}.$$

Solving the recursion in the eigenvalue formulation gives an alternating product for the tree numbers:

$$\tau_d = \prod_{i=0}^{d} \pi_d^{(-1)^{d-i}}.$$

Example 7.6.10. As in Example 7.6.5, let Δ be the boundary complex of a polytope with m facets. We have already seen that Δ has one spanning tree for each facet and so Δ has m trees total.

Let us confirm this enumeration for the example of the tetrahedron via Theorem 7.6.9. The Laplacian L_2 is shown in Example 7.3.4. The one skeleton $\Delta_{(1)}$ is the graph K_4.

Fix the spanning tree Υ of $\Delta_{(1)}$ consisting of all edges that contain the vertex 1.

Then the reduced Laplacian is:

$$
L_\Upsilon = \begin{array}{c} \\ 23 \\ 24 \\ 34 \end{array}
\begin{array}{c} \begin{array}{ccc} 23 & 24 & 34 \end{array} \\
\left(\begin{array}{ccc}
2 & -1 & 1 \\
-1 & 2 & -1 \\
1 & -1 & 2
\end{array} \right). \end{array}
$$

There are no torsion factors in this example and

$$
\tau_2 = \det L_\Upsilon = 4.
$$

Alternatively,
The non-zero eigenvalues of L_2 are: $(4, 4, 4)$.
The non-zero eigenvalues of L_1 are: $(4, 4, 4)$.
The non-zero eigenvalues of L_0 are: (4).

Therefore the number of two-dimensional spanning trees can also be computed as

$$
\tau_2 = \frac{\pi_2 \cdot \pi_0}{\pi_1} = \frac{4 \cdot 4 \cdot 4 \cdot 4}{4 \cdot 4 \cdot 4} = 4.
$$

The Cellular Matrix-Tree Theorem holds in greater generality than stated here. The requirement that $H_k(\Sigma; \mathbb{Z}) = 0$ for all $k < d$ yields cleaner formulas for the tree numbers and is sufficiently general for our purposes. For the remainder of the chapter, we will assume all complexes have $H_k(\Sigma; \mathbb{Z}) = 0$ for all $k < d$.

Example 7.6.11 (Kalai's formula). [Kal83] Let $\Delta_{(n,d)}$ be the complete complex of dimension d on n vertices consisting of all possible d-dimensional faces on n vertices, i.e. every collection of $d+1$ vertices is a facet of $\Delta_{(n,d)}$. Then

$$
\tau_d(\Delta_{(n,d)}) = n^{\binom{n-2}{d}}.
$$

Example 7.6.12. Consider the simplicial complex RP^2 shown in Figure 7.9 which is a triangulation of the real projective plane consisting of 6 vertices and 10 triangles. This complex is a 2-dimensional tree. It is the smallest example of a simplicial tree with torsion: $H_1(RP^2; \mathbb{Z}) \cong \mathbb{Z}/2\mathbb{Z}$.

The tree number has only one summand, but the one tree is given the weight 4.

$$
\tau_2(RP^2) = 2^2 = 4.
$$

The triangulation RP^2 is also a spanning tree of the complete complex $\Delta_{(6,2)}$ of Example 7.6.11. The contribution to Kalai's formula is again equal to 4 and not equal to 1. In contrast to Kalai's formula, the exact number of spanning trees of $\Delta_{(n,k)}$ (the unweighted count) does not seem to be nice; there is no known closed formula.

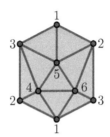

FIGURE 7.9: A triangulation of the real projective plane.

7.7 Sandpile groups

Having now introduced the theory of higher-dimensional trees, we revisit the sandpile groups of cell complexes. Recall that for a d-dimensional complex Δ with d-dimensional Laplacian L,

$$\mathcal{S}(\Delta) = \ker \partial_{d-1} / \operatorname{im} L.$$

As in the graphical case, the sandpile group can be seen in terms of a reduced combinatorial Laplacian. In the one-dimensional case, the Laplacian is reduced by a single row and column corresponding to a sink vertex. The sink vertex can be seen as a spanning tree of the 0-skeleton of the graph.

Definition 7.7.1. For a d-dimensional complex Δ, a *sink* is a fixed simplicial tree of dimension $d-1$.

Theorem 7.7.2 ([DKM13]).
Let Δ be a d-dimensional complex such that $H_k(\Delta; \mathbb{Z}) = 0$ for all $k < d$. Let Υ be a sink of Δ such that $H_{d-2}(\Upsilon; \mathbb{Z}) = 0$. Let Θ be the set of faces of $\Delta_{(n-1)}$ not in Υ, $\Theta = \Delta_{(n-1)} \setminus \Upsilon$.
Then

$$\mathcal{S}(\Delta) \cong \operatorname{coker}(L_\Upsilon) \cong \mathbb{Z}^\Theta / \operatorname{im} L_\Upsilon,$$

where L_Υ is the Laplacian restricted to faces in Θ.

Corollary 7.7.3. *Under the conditions of Theorem 7.7.2, the order of the sandpile group is the torsion-weighted number of spanning trees:*

$$|\mathcal{S}(\Delta)| = \tau_d.$$

Therefore, the number of equivalence classes of the flow firing relationship is equal to the torsion-weighted number of spanning trees.

Example 7.7.4. For the complex RP^2, there is only one spanning tree, but

$$|\mathcal{S}(RP^2)| = 4.$$

There are indeed four distinct firing classes of flow configurations on the edges of RP^2.

The connection between spanning trees and elements of the sandpile group was the starting point for the many combinatorial connections of Chapter 3. In higher dimensions, there are not yet many extensions of the combinatorial theory. For example, there are no general bijections between firing classes and spanning trees.

We mention one result in this direction. Recall from Chapter 4 Section 4.7.3 the circuit–cocircuit reorientation torsor. In the general case of this construction, Backman, Baker and Yuen [BBY17] work with the class of regular matroids. They set up a family of bijections between bases of the matroid and circuit–cocircuit reorientation classes. For a regular simplicial matroid, this gives a bijection between spanning trees and circuit–cocircuit reorientation classes. We will revisit this circle of ideas at the end of Section 7.9.1.

As illustrated in Theorem 7.7.2, a conservative flow configuration $\mathbf{c}$ can be given either by specifying the value of $\mathbf{c}$ at all faces of dimension $d - 1$ or by specifying the values of $\mathbf{c}$ at all non-sink faces of dimension $d - 1$. In the second case, the values of $\mathbf{c}$ on sink faces are uniquely determined by the conservation condition.

Example 7.7.5. Let Δ be the complex consisting of three triangles from Example 7.4.1. Let the sink Υ be the 1-dimensional spanning tree consisting of all edges that contain the vertex v_1. Suppose we have an initial configuration on the edges of $\Theta = \Delta_{(1)} \setminus \Upsilon$, as shown in the top complex of Figure 7.10.

There are three non-sink edges. With the values given: $(2, 5, 4)$ and

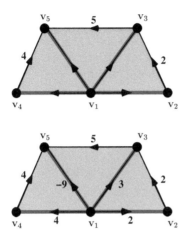

FIGURE 7.10: Flow on a simplicial complex with sink.

the requirement that the inflow(v) − outflow(v) equal zero at each vertex v, the full configuration is uniquely determined to be

$$(2, 3, 4, -9, 2, 5, 4).$$

Firing edge $v_1 v_2$ results in the configuration: $(1, 4, 4, -9, 1, 5, 4)$.
Firing edge $v_1 v_3$ results in the configuration: $(2, 2, 4, -8, 2, 4, 4)$.

In this example, the complex Δ itself is a two-dimensional tree. All conservative flows are firing equivalent.

In general, what does it mean to normalize a configuration **c** to be in the kernel of ∂_d?

- For $d = 1$ the total sum of chips is constant and equal to 0.

- For $d = 2$ inflow(v) = outflow(v) at each vertex v. (There is no accumulation or depletion at any vertex.)

- For $d = 3$ flow is on 2-cells (thought of as circulation) and there is no accumulation or depletion at any edge.

$$\vdots$$

7.7.1 Precise forms of sandpile groups

The sandpile group of a complex is a finite abelian group presented as the integer cokernel of a matrix. Thus the theory of invariant factors and the Smith normal form, discussed in Chapter 4, continues to apply to higher-dimensional sandpile groups. As with the graphical case, understanding the precise form, i.e. the invariant factors, of special classes of complexes proves to be quite difficult.

The following can be seen as a higher-dimensional analog of the results of Lorenzini and Merris who proved that the sandpile group of an n-cycle is cyclic; see Chapter 4 Section 4.5.1.

Theorem 7.7.6 ([DKM15]). *Let Δ be a d-dimensional simplicial sphere with n facets. Then*

$$\mathcal{S}(\Delta) \cong \mathbb{Z}/n\mathbb{Z}.$$

Theorem 7.7.7 (Kalai). *Let $\Delta_{(n,d)}$ be the complete complex on n vertices which consists of all possible faces of dimension d. Then*

$$\mathcal{S}(\Delta_{(n,d)}) \cong \mathbb{Z}/n\mathbb{Z}^{\binom{n-2}{d}}.$$

Theorem 7.7.7 is the chip-firing equivalent of Kalai's formula (Example 7.6.11) for the tree number of the complete complex.

Many questions remain about higher sandpile groups. What does a random sandpile group look like for cell complexes of a fixed dimension? For example, can the results of Wood be extended to higher dimensions? Combinatorial Laplacians differ from graph Laplacians in important ways. For example, they are not M-matrices.

7.8 Cuts and flows

Section 4.6 of Chapter 4 presented the sandpile group in terms of both the cut and flow lattices of a graph. For a graph, the sandpile group is isomorphic to the discriminant groups of the cut and flow lattices and the cutflow group:

$$\mathcal{C}^\sharp/\mathcal{C}, \quad \mathcal{F}^\sharp/\mathcal{F}, \quad \mathbb{Z}^{|E|}/(\mathcal{C} \oplus \mathcal{F}).$$

The lattices and cut and flow spaces are constructed in terms of the oriented incidence matrix of the graph. Following these constructions, using higher-order boundary maps, leads to a higher-dimensional theory of cuts and flows as well as a collection of algebraic invariants associated to a cell complex. This section is based on the work in [DKM15].

Definition 7.8.1. The *cut and flow spaces* and *cut and flow lattices* of a cell complex Σ are

$$\mathrm{Cut}(\Sigma) = \mathrm{im}_\mathbb{R}\, \partial^T, \qquad\qquad \mathrm{Flow}(\Sigma) = \ker_\mathbb{R} \partial,$$
$$\mathcal{C}(\Sigma) = \mathrm{im}_\mathbb{Z}\, \partial^T, \qquad\qquad \mathcal{F}(\Sigma) = \ker_\mathbb{Z} \partial,$$

where ∂ and ∂^T are the top cellular boundary and coboundary maps of Σ.

The definitions of cuts and flows are identical to the 1-dimensional case. In topological terms, cut- and flow-vectors are cellular coboundaries and cycles, respectively. Equivalently, the vectors in $\mathrm{Cut}(\Sigma)$ support sets of facets whose deletion increases the codimension-1 Betti number, and the vectors in $\mathrm{Flow}(\Sigma)$ support nontrivial rational homology classes.

In the language of matroid theory, cuts and flows correspond to cocircuits and circuits of the cellular matroid $M(\Sigma)$.

$$\text{cuts} \longleftrightarrow \text{cocircuits}$$
$$\text{flows} \longleftrightarrow \text{circuits}$$

Su and Wagner [SW10] define the cut and flow lattices for any regular matroid M via the integer span of the image and kernel of a totally unimodular representation of M. Cellular matroids are not necessarily regular matroids, a complex with torsion does not have a representation by a totally unimodular matrix.

Example 7.8.2. Let Σ be the equatorial bipyramid; the 2-dimensional complex consisting of the boundaries of two tetrahedra glued together along a single facet as shown on the right in Figure 7.11.

The collection of facets $\{123, 125, 234\}$ is a cut of Σ. On the left in Figure 7.11, these cut facets have been deleted. The remaining complex has a "hole," as if it had been pierced. It is contractible to a circle. The cut is a cocircuit of $M(\Sigma)$.

The collection of facets $\{123, 124, 134, 234\}$ is a flow of Σ. On the right in Figure 7.11, these flow facets have been marked with a curved arrow; they form a circuit in $M(\Sigma)$.

Higher-dimensional cuts and flows can be formed via fundamental

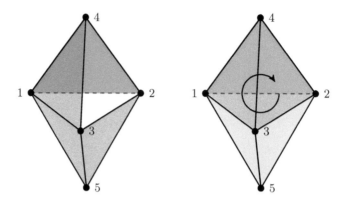

FIGURE 7.11: On the left, the three faces of the cut $\{123, 125, 234\}$ have been removed. On the right, the top four faces $\{123, 124, 134, 234\}$ form a flow.

circuits and fundamental bonds as in Chapter 4 Section 4.6. Such constructions are purely matroidal. Some care, however, must be taken to form *bases* of the cut and flow spaces and lattices. In higher dimensions, it is not sufficient to consider characteristic vectors of fundamental cuts and bonds, some arithmetic information must also be used; see [DKM15].

In the higher-dimensional setting, the groups $\mathcal{C}^\sharp/\mathcal{C}$ and $\mathcal{F}^\sharp/\mathcal{F}$ are not necessarily isomorphic to each other, precisely because of the presence of torsion. In order to see their relationship to the sandpile group, we first define the co-sandpile group.

Definition 7.8.3. Given a d-dimensional complex Σ such that $H_k(\Sigma; \mathbb{Z}) = 0$ for all $k < d$, an *acyclization* of Σ is a $(d+1)$-dimensional complex Ω such that $\Omega_{(d)} = \Sigma$ and $H_{d+1}(\Omega; \mathbb{Z}) = H_d(\Omega; \mathbb{Z}) = 0$.

Algebraically, this construction corresponds to finding an integral basis for $\ker \partial_d(\Sigma)$ and declaring its elements to be the columns of $\partial_{d+1}(\Omega)$.

Definition 7.8.4. The *co-sandpile group* $\mathcal{S}^*(\Sigma)$ is

$$\mathcal{S}^*(\Sigma) = C_{d+1}(\Omega; \mathbb{Z}) \, / \, \operatorname{im} \partial_{d+1}^T \partial_{d+1}.$$

Although not immediate, the group $\mathcal{S}^*(\Sigma)$ is independent of the choice of the acyclization Ω.

Theorem 7.8.5 ([DKM15]).
Let Σ be a cell complex of dimension d. Then we have the short exact

sequences:

$$0 \to \mathbb{Z}^n/(\mathcal{C} \oplus \mathcal{F}) \to \mathcal{S}(\Sigma) \cong \mathcal{C}^\sharp/\mathcal{C} \to \mathbf{T}(H_{d-1}(\Sigma; \mathbb{Z})) \to 0, \quad and$$
$$0 \to \mathbf{T}(H_{d-1}(\Sigma; \mathbb{Z})) \to \mathbb{Z}^n/(\mathcal{C} \oplus \mathcal{F}) \to \mathcal{S}^*(\Sigma) \cong \mathcal{F}^\sharp/\mathcal{F} \to 0,$$

where $\mathbf{T}$ *denotes the torsion summand.*

Corollary 7.8.6.
The sizes of the groups of Theorem 7.8.5 are given by:

$$\begin{aligned}
|\mathcal{C}^\sharp/\mathcal{C}| = |\mathcal{S}(\Sigma)| &= \tau(\Sigma) &= \tau^*(\Sigma) \cdot \mathbf{t}^2, \\
|\mathbb{Z}^n/(\mathcal{C} \oplus \mathcal{F})| &= \tau(\Sigma)/\mathbf{t} &= \tau^*(\Sigma) \cdot \mathbf{t}, \\
|\mathcal{F}^\sharp/\mathcal{F}| = |\mathcal{S}^*(\Sigma)| &= \tau(\Sigma)/\mathbf{t}^2 &= \tau^*(\Sigma),
\end{aligned}$$

where $\mathbf{t} = |\mathbf{T}(H_{d-1}(\Sigma; \mathbb{Z}))|$ *and* $\tau(\Sigma)$ *and* $\tau^*(\Sigma)$ *are the weighted enumerators*

$$\tau(\Sigma) = \sum_\Upsilon |\mathbf{T}(H_{d-1}(\Upsilon; \mathbb{Z}))|^2, \quad \tau^*(\Sigma) = \sum_\Upsilon |\mathbf{T}(H_d(\Omega, \Upsilon; \mathbb{Z}))|^2,$$

where both sums run over all cellular spanning forests $\Upsilon \subseteq \Sigma$ *and* Ω *is an acyclization of* Υ.

Example 7.8.7. Let Σ be the triangulation of $\mathbb{R}P^2$ as in Example 7.6.12. We have:

$$\mathcal{S}(\Sigma) \cong \mathcal{C}^\sharp/\mathcal{C} = \mathbb{Z}/4\mathbb{Z}$$
$$\mathbb{Z}^n/(\mathcal{C} \oplus \mathcal{F}) = \mathbb{Z}/2\mathbb{Z}$$
$$\mathcal{S}^*(\Sigma) \cong \mathcal{F}^\sharp/\mathcal{F} = 0.$$

Corollary 7.8.8. *If* $H_{d-1}(\Sigma; \mathbb{Z})$ *is torsion-free, then the groups of Theorem 7.8.5 are all isomorphic to each other:*

$$\mathcal{S}(\Sigma) \cong \mathcal{S}^*(\Sigma) \cong \mathcal{C}^\sharp/\mathcal{C} \cong \mathcal{F}^\sharp/\mathcal{F} \cong \mathbb{Z}^n/(\mathcal{C} \oplus \mathcal{F}).$$

7.9 Stability

What does it mean for a configuration to be stable in higher dimensions? For a flow configuration, we have seen that the sign of an entry corresponds to a direction of flow. We have also seen that firing an edge can potentially decrease the amount of flow on neighboring edges. As such, local non-negativity is no longer a natural stopping condition. We might think to use the absolute value of flow as a stopping condition, as in one of our first examples on the cubical grid. While this is natural in some situations, it does not in general lead to the good behavior we would expect from a chip-firing process. For example, in the cubical example this rule gave stabilization but not confluence. Moreover, even for a finite complex with a sink, it does not yield a system of representatives for the sandpile group.

7.9.1 M-pairings

The combinatorial Laplacian is not an M-matrix, it is not even a Z-matrix as it has positive off-diagonal entries (see Chapter 6 and Section 6.1). The reduced combinatorial Laplacian is a non-singular integer matrix. By choosing a suitably sized M-matrix, we can use the construction of M-matrix pairings from Chapter 6 Section 6.7 to define critical and superstable configurations that are systems of representatives for the sandpile group.

Example 7.9.1 ([GK16]). Consider our running example of the boundary complex of a tetrahedron, Δ. Let Υ be the spanning tree of $\Delta_{(1)}$ consisting of all edges containing the vertex 1. As in Example 7.6.10, the reduced Laplacian is:

$$
L_\Upsilon = \begin{array}{c} \\ 23 \\ 24 \\ 34 \end{array} \begin{array}{ccc} 23 & 24 & 34 \\ \left(\begin{array}{ccc} 2 & -1 & 1 \\ -1 & 2 & -1 \\ 1 & -1 & 2 \end{array} \right). \end{array}
$$

This matrix also appeared in Example 6.7.3 of Chapter 6.

Pairing L_Υ with the M-matrix:

$$
M = \begin{pmatrix} 3 & -1 & -1 \\ -1 & 3 & -1 \\ -1 & -1 & 3 \end{pmatrix},
$$

yielded the critical configurations:

$$
\begin{pmatrix} 4 \\ -1 \\ 4 \end{pmatrix}, \begin{pmatrix} 4 \\ 0 \\ 4 \end{pmatrix}, \begin{pmatrix} 5 \\ 0 \\ 5 \end{pmatrix}, \begin{pmatrix} 5 \\ -1 \\ 5 \end{pmatrix}
$$

and the superstable configurations:

$$
\begin{pmatrix} 1 \\ 0 \\ 1 \end{pmatrix}, \begin{pmatrix} 1 \\ 1 \\ 1 \end{pmatrix}, \begin{pmatrix} 0 \\ 0 \\ 0 \end{pmatrix}, \begin{pmatrix} 2 \\ 1 \\ 2 \end{pmatrix}.
$$

We now interpret these configurations as flows on the edges of the tetrahedron. Figure 7.12 shows the three non-zero superstable configurations. The edges of the sink are dashed and their flow values suppressed. The sink values could be recovered via the conservation requirement. They can be thought of as the stable configurations if multiple edges can reroute multiple flows at once.

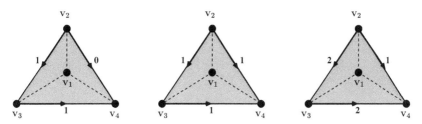

FIGURE 7.12: The three non-zero superstable configurations for the pairing (L_Υ, M).

If we use the identity matrix as our chosen M-matrix, then the critical and superstable configurations coincide and are:

$$\begin{pmatrix} 1 \\ 0 \\ 1 \end{pmatrix}, \begin{pmatrix} 0 \\ 1 \\ 0 \end{pmatrix}, \begin{pmatrix} 0 \\ 0 \\ 0 \end{pmatrix}, \begin{pmatrix} 2 \\ -1 \\ 2 \end{pmatrix}.$$

Figure 7.13 shows the three non-zero superstable configurations. Two of the four configurations are the same as the earlier pairing.

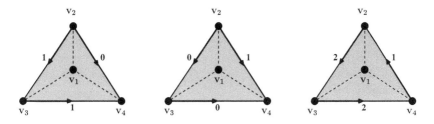

FIGURE 7.13: The three non-zero superstable (and critical) configurations for the pairing (L_Υ, I).

One narrative for this system is as a model of traffic flow; see [GK16]. The edges represent streets and the vertices are intersections. Sink edges can be thought of as major boulevards that can handle large amounts of traffic. As local streets become congested, traffic diverts to neighboring streets. Informally, the system stabilizes when as much traffic as possible has been diverted to the major boulevards.

The third image of Figure 7.13 is a solution in which the directionality of street 24 has been reversed.

The theory of M-pairings from Chapter 6, Section 6.7 shows that the critical and superstable configurations form systems of representatives for the sandpile group.

In the spirit of Chapter 3, one would like to combinatorially connect these families of stable configurations and higher-dimensional trees (equiv. bases of the cellular matroid). The regular case could support bijective correspondence.

Let Σ be a finite cell-complex satisfying $\beta_{d-1}(\Sigma) = 0$. Let L be the top-dimensional Laplacian of Σ. Suppose further that the top boundary map $\partial(\Sigma)$ is a totally unimodular matrix, so that $M(\Sigma)$ is a regular matroid. Let M be a M-matrix.

Then the following are the same:

1. The number of firing equivalence classes $|\operatorname{coker}(L_\Upsilon)|$.

2. The number of critical configurations of the pair (L_Υ, M).

3. The number of superstable configurations of the pair (L_Υ, M).

4. The number of bases of the cellular matroid of $M(\Delta)$.

5. The determinant $\det(L_\Upsilon)$.

Thus there necessarily exist bijections between the superstable configurations, critical configurations, and the bases of the cellular matroid.

There are no known explicit bijections between any pair of the three. There is no known concept of duality in this general setting so even a bijection between critical and superstable configurations is in fact an open question.

In general, the number of critical (or superstable) configurations is the torsion weighted tree count, which can be larger than the number of bases of the cellular matroid.

One approach here would be to define a (not necessarily bijective) map from the sandpile group to the bases of the cellular matroid

$$\mathcal{S}(\Sigma) \to \mathcal{B}(M(\Sigma))$$

associating possibly many critical configurations to a single spanning tree.

7.10   Exercises

Exercise 7.10.1. *Simulate higher-dimensional chip-firing. Some possibilities:*

1. *Chip-fire from an initial circulation on non-grid planar lattices such as the honeycomb.*

2. *Chip-fire on a triangulation of the sphere.*

3. *Chip-fire on the two-dimensional faces of the three-dimensional grid.*

4. *Chip-fire on the two-dimensional grid with non-zero initial background.*

Exercise 7.10.2. *Consider the cubical example of Section 7.1. Show that there are finite initial configurations that do not stabilize under the chip-firing rules of that example.*

Exercise 7.10.3. *Consider the triangulation of RP^2 in Figure 7.9. The size of the sandpile group is $4 = 2^2$. Find four non firing-equivalent flow configurations of the complex.*

Exercise 7.10.4.

1. *Prove the equivalences of Definition 7.6.1.*

2. *Prove the equivalences of Definition 7.6.3.*

Exercise 7.10.5.

1. *Prove the statement of Theorem 7.7.6 that the sandpile group of a simplicial sphere with n facets is $\mathbb{Z}/n\mathbb{Z}$.*

2. *Extend the result to all psuedomanifolds with n facets.*

Exercise 7.10.6. *Compute the sandpile group of the boundary of the cone over RP^2.*

Exercise 7.10.7. *Let $\Delta_{(5,2)}$ be the complete two-dimensional complex on 5 vertices. Compute the eigenvalues of the combinatorial Laplacians of $\Delta_{(5,2)}$.*

Exercise 7.10.8. *List all two-dimensional simplicial spanning trees of the equatorial bipyramid.*

Exercise 7.10.9. *Let Δ be the boundary of the equatorial bipyramid.*

    1. Compute the boundary maps and Laplacian of Δ as in Example 7.4.1.

    2. Fix an initial configuration on Δ and perform firing moves.

    3. Fix a spanning tree of the one-skeleton of Δ as a sink. As in Example 7.7.5, fix an initial configuration on Δ and perform firing moves.

In both cases, explore the differences between firing on edges and firing on vertices.

Exercise 7.10.10. *Prove that for a complex Σ, the co-sandpile group $S^*(\Sigma)$ is independent of the choice of acyclization of Σ.*

Exercise 7.10.11. *Compute the five groups of Corollary 7.8.8 for the equatorial bipyramid. Verify that they are indeed isomorphic.*

Exercise 7.10.12. *Let Δ be the equatorial bipyramid.*
    Fix a spanning tree Υ of the one skeleton of Δ. Fix a 5×5 M-matrix M. Compute the critical and superstable configurations for the pair (Δ_Υ, M).

Exercise 7.10.13. *Let $G = K_4 \setminus e$ with reduced Laplacian L_q. Consider chip-firing with a pair (L_q, M) for various M-matrices. How do the critical and superstable configurations change with various choices of M mediating between L_q and I?*

Exercise 7.10.14. *Consider a chip configuration on the edges of a two-dimensional complex Δ as a formal linear combination of the edges. Show that firing an edge can be interpreted as adding or subtracting a boundary, i.e. an element of $\mathrm{im}(\partial_2(\Delta))$.*

Exercise 7.10.15. *(Open)*
For a complex Δ with torsion trees, define a natural many-to-one map from $S(\Delta)$ to $T(\Delta)$.

Chapter 8

Divisors

This chapter looks at chip-firing from a perspective motivated by algebraic geometry. From this perspective, finite graphs provide a discrete model for Riemann surfaces and chip configurations play the role of divisors on curves.

Bacher, de la Harpe, and Nagnibeda consider the analogy between graphs and curves in [BdlHN97]. In the context of studying the lattices of cuts and flows (see Section 4.6) they define the Picard group and Jacobian of a graph. Bacher, de la Harpe, and Nagnibeda further explore an Abel–Jacobi theory for graphs and ask about other extensions such as a Torelli's Theorem for graphs. Continuing in the same direction, Kotani and Sunada [KS00] study the Jacobian and Albanese tori of graphs. Biggs [Big97] concentrates on the role of the Laplacian and considers the analogy to curves in relation to algebraic potential theory on graphs.

The idea of divisors on graphs gained considerable attention with the work of Baker and Norine [BN07]. Baker and Norine introduce the *rank* of a chip configuration and prove a Riemann–Roch Theorem for graphs (Theorem 8.3.1). Once suitable definitions have been made, the statement of the Theorem is identical to the classical case:
For all divisors D and canonical divisor K,

$$r(D) - r(K - D) = \deg(D) + 1 - g,$$

where r is the rank function on divisors.

Lorenzini similarly considered the analogy between graphs and curves in [Lor89]. Lorenzini's perspective was motivated by arithmetic geometry. In particular, he introduced arithmetical graphs in the context of degenerations of curves. Here the sandpile group is referred to as the group of components. The second half of the chapter focuses on this arithmetic geometry perspective, including Lorenzini's Riemann–Roch theory for lattices and two-variable zeta-function.

8.1   Divisors on curves

The chip-firing theory of divisors on curves returns us to the setting of finite graphs. While the motivating narrative is from algebraic geometry, the setting and tools are graph theoretic. We first set up a dictionary of terminology for divisors.

By regarding a graph as an analog of a Riemann surface, chip configurations are interpreted as divisors on curves. Through this lens, chip configurations are not presented as integer vectors but formal sums of vertices.

Definition 8.1.1. For a graph G on vertex set V, a *divisor* D on G is any formal sum of vertices:

$$D = \sum_{v \in V} D(v)\, v,$$

where $D(v) \in \mathbb{Z}$. The collection of all divisors on G is the free abelian group on the vertices of G, the *group of divisors* and denoted $\mathrm{Div}(G)$.

Linear equivalence takes the place of chip-firing equivalence.

Definition 8.1.2. Two divisors C and D on G are called *linearly equivalent*, $C \sim D$, if the difference in coefficients of C and D is in the image of the graph Laplacian.

The equivalence classes of the chip-firing relation are now called the divisor classes.

Definition 8.1.3. The *divisor class* of D is the collection of all divisors linearly equivalent to D and denoted $[D]$:

$$[D] = \{D' \in \mathrm{Div}(G) \mid D' \sim D\}.$$

In an unfortunate conflict of terminology with graph theory, the sum of all coefficients of a divisor is the degree of a divisor.

Definition 8.1.4. For a divisor D of G, the *degree* of D is

$$\deg(D) = \sum_{v \in V} D(v).$$

Note that in our new terminology we have that linearly equivalent divisors have the same degree.

$$C \sim D \implies \deg(C) = \deg(D).$$

Let $\mathrm{Div}^k(G)$ denote the collection of all divisors of degree k. We will work primarily with $\mathrm{Div}^0(G)$, the divisors of degree 0.

Definition 8.1.5. A divisor on G is *effective* if $D(v) \geq 0$ for all v.

Definition 8.1.6. For a divisor D on a graph G, the *linear system of D* is the collection of all effective divisors linearly equivalent to D:

$$|D| = \{E \in \mathrm{Div}(G) \mid E \geq 0, E \sim D\}.$$

The most important new concept that we will be working with in this chapter is the rank of a divisor.

Definition 8.1.7. The *rank* of a divisor $r(D)$ is defined constructively as follows:

- If D is not equivalent to any effective divisor then

$$r(D) = -1.$$

- $r(D) \geq k$ if and only if for any removal of k chips from D, the resulting divisor is still equivalent to an effective divisor.

Definition 8.1.8. The *canonical divisor K_G* on G is

$$K_G = \sum_v (\deg(v) - 2)v,$$

where $\deg(v)$ is the usual graphical degree of the vertex v.

The degree of the canonical divisor is $\deg(K_G) = 2g - 2$, where

$$g = E - V + 1$$

is the *genus* or cyclomatic number of G as seen for example in Chapter 3 in Theorem 2.3.6. The genus g is the upper bound on the level of a configuration; equivalently it is the degree of the critical polynomial; see Section 3.2

Those familiar with the theory of divisors on curves will recognize the terminology setup here. For most concepts, the definitions carry over word for word from the case of curves. The rank, whose definition is not familiar, can be thought of as the dimension of the complete linear system of a divisor – it will play this role in the Riemann–Roch Theorem.

As a final piece of terminology, we redefine superstable configurations.

Definition 8.1.9. For a graph G with a sink vertex q, a divisor D on G is called *q-reduced* if

1. $D(v) \geq 0 \ \forall v \neq q$.

2. For any $A \subset V \setminus q$, the divisor resulting from the cluster-fire of A is not effective.

In summary, we have the following correspondences, we include the sandpile group for completeness. It is the focus of the next section.

Curves	Graphs
Divisor D	Chip configuration $\mathbf{c}$
$\deg(D)$	$\mathrm{wt}(\mathbf{c})$
Canonical K	$\mathbf{c}_{\max} - 1$
Effective D	$\mathbf{c} \geq 0$
Linearly equivalent	Firing equivalent
Divisor class	Firing class
q-reduced	Superstable
Picard group / Jacobian	Sandpile group

8.2   The Picard group and Abel–Jacobi theory

For a graph with n vertices, a chip configuration is an integer vector where the ith entry is interpreted as the number of chips at the ith vertex. A divisor is a formal linear combination of vertices where the number of chips at the ith vertex is the coefficient of the ith vertex. There is yet another way to encode the number of chips at each vertex. Define

$$\mathcal{M}(G) = \mathrm{Hom}(V; \mathbb{Z}).$$

$\mathcal{M}(G)$ can be thought of in analogy to the meromorphic functions of a Riemann surface.

The Laplacian can then be interpreted as an operator from functions to divisors

$$L : \mathcal{M}(G) \to \text{Div}(G).$$

Recall that the group of divisors $\text{Div}(G)$ is the free abelian group on the vertices of G. Define the subgroup of *principal* divisors $\text{Prin}(G) \subset \text{Div}(G)$ as those divisors in the image of the Laplacian:

$$\text{Prin}(G) = L(\mathcal{M}(G)).$$

Two divisors are linearly equivalent if their difference is a principal divisor. The degree of a principal divisor is zero, thus $\text{Prin}(G) \subset \text{Div}^0(G)$.

Baker and Norine [BN07] define the *Jacobian* of a graph as the quotient of the degree zero divisors by the principal divisors.

Bacher, de la Harpe, and Nagnibeda [BdlHN97] define the *Picard* group of a graph as this quotient:

$$\text{Pic}(G) = \text{Div}^0(G)/\text{Prin}(G).$$

Regardless of name, we recognize the quotient as the sandpile group

$$\mathcal{S}(G) = \ker \partial_0(G)/\text{im}\, L(G).$$

Bacher, de la Harpe, and Nagnibeda define an alternative construction for the Jacobian of a graph. As one might anticipate, the two groups are isomorphic. The construction is as follows.

Let $C^1(G; \mathbb{R})$ be the space of real valued functions on oriented edges of G; see Chapter 7. The space $C^1(G; \mathbb{R})$ has an orthogonal Hodge decomposition:

$$C^1(G; \mathbb{R}) = \ker(L_1) \oplus \text{im}(\partial_1),$$

where L_1 is the graph Laplacian of G and ∂_1 is the oriented incidence matrix of G. The elements of the kernel of the Laplacian are thought of as harmonic one forms on G. Define

$$\Gamma^1(G) = C^1(G; \mathbb{Z}) \cap \ker(L_1),$$

which we recognize as the lattice of integral flows of G; see Section 4.6 of Chapter 4. The Jacobian of a graph G is then defined as the discriminant group of flows:

$$\text{Jac}(G) = \Gamma^1(G)^\sharp/\Gamma^1(G),$$

thought of in analogy to $H^0(X, \Omega)^*/H_1(X, \mathbb{Z})$, the dual of the space of holomorphic 1-forms quotient by the lattice $H_1(X, \mathbb{Z})$.

Already in Chapter 4, we saw that the discriminant group of the flow lattice is isomorphic to the sandpile group. In that context, we worked explicitly with the projection matrices for lattices. In the current context, the isomorphism between the Jacobian and the Picard group is part of an Abel–Jacobi theory for graphs.

Definition 8.2.1. A map $\phi : G \to A$ from the vertices of G to an abelian group A is *harmonic* if at each vertex $\phi(v)$ is equal to the average of the values at the neighbors of v:

$$\phi(v) = \frac{\sum_{\{vw\} \in E(G)} \phi(w)}{\deg(v)}.$$

Given a graph G, fix a basepoint of G. We will think of the basepoint as the sink of G.

Definition 8.2.2. For a graph G with basepoint s, define the map ρ_s, by

$$\rho_s : G \to \mathrm{Div}^0(G)/\mathrm{Prin}(G)$$
$$\rho_s(v) = [(v) - (s)].$$

The map ρ_s sends a vertex v to the divisor class that contains the divisor with coefficient 1 at v and -1 at s. Note that the map ρ_s is harmonic, it maps s to 0, and the image of ρ_s generates all of $\mathrm{Pic}(G)$.

The map ρ_s has the following universal property [BdlHN97, Section 3]. Let A be an abelian group. If $\phi : G \to A$ is a harmonic map with $\phi(s) = 0$, then there is a unique $\psi : \mathrm{Pic}(G) \to A$ such that $\phi = \psi \circ \rho_s$:

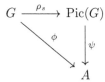

We will use this universality to relate the Jacobian to the Picard group.

Definition 8.2.3. For a graph G with basepoint s, define the map A_s by:

$$A_s : G \to \Gamma^1(G)^\sharp / \Gamma^1(G)$$
$$A_s(v) = [s \to v],$$

where $[s \to v]$ denotes the class of oriented paths from s to v.

Note that the map A_s is well defined because any closed path from s to s is in $\Gamma^1(G)$. Bacher, de la Harpe, and Nagnibeda call A_s the *Abel–Jacobi* map from a graph to its Jacobian. The map A_s is harmonic with $A_s(s) = 0$.

Thus, the universality property gives a unique isomorphism

$$\psi : \text{Pic}(G) \;\rightarrow\; \text{Jac}(G).$$

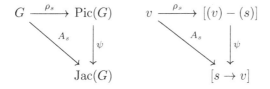

8.3   Riemann–Roch Theorems

In this section we prove the Riemann–Roch Theorem for graphs. The Theorem is due to Baker and Norine [BN07]. The result sparked new interest in chip-firing processes, especially from a more algebraic perspective.

Theorem 8.3.1 (Riemann–Roch for Graphs, [BN07])**.** *Let G be a finite graph, D a divisor on G and K the canonical divisor on G, then*

$$r(D) - r(K - D) = \deg(D) + 1 - g.$$

Before proving Theorem 8.3.1, we take a closer look at the rank function $r(D)$.

8.3.1   The rank function

The concept of the rank of a divisor is closely related to Biggs' dollar game [Big99a]. In Biggs' study of chip-firing processes, a graph represents a network of economies. The value at a particular site is interpreted as a dollar amount, where negative values represent debt. Firing a vertex is a *lending move* – a vertex gives one dollar to each neighbor. There is also a *borrowing move* – a fixed vertex takes one dollar from each neighbor. The borrowing move results in adding a row of the Laplacian as opposed to the usual lending move which results in subtracting a row of the Laplacian.

Consider the following question.

The dollar game: Given a graph G and an initial set of values on the vertices of G can you get everyone out of debt using only lending and borrowing moves?

In the language of this chapter:

The dollar game: Given a divisor D, is D linearly equivalent to an effective divisor?

The rank records how robust a configuration is to this game. Given a divisor D:

- If the answer is no (if everyone cannot simultaneously get out of debt) then the rank is negative,

$$r(D) = -1.$$

- If the answer is yes (if everyone can get out of debt) then the rank function can be thought of in terms of an adversarial player:

  If any 1 chip is removed from D, can everyone still get out of debt?

  If not, then the rank is equal to 0. If yes, then the problem becomes:

  If any 2 chips are removed from D, can you still get everyone out of debt? If not, then the rank is equal to 1. If yes, ...

Kiss and Tóthmérész show the following complexity result for determining the rank.

Proposition 8.3.2 ([KT15])**.** *Computing the rank of a divisor on a (simple) graph is NP-Hard.*

In order to analyze the computation, they use the following reformulation of the rank: Given a configuration $\mathbf{c}$, the rank is the minimal number of chips that can be added to $\mathbf{c}$ in order to form a configuration that does not stabilize.

8.3.2 Proof

The Riemann–Roch Theorem for graphs is originally due to Baker and Norine. We will present an alternative proof which is due to Cori and Le Borgne [CLB16].

The proof will use the next result about when configurations can get out of debt. It is a theorem of alternatives, in the spirit of an integer Farkas Lemma. The result considers divisors of the form $\text{outdeg}(\mathcal{O}) - \mathbf{1}$ for an acyclic orientation $\mathcal{O}$. We have considered such configurations before. In particular, Theorem 3.5.1 gives a bijection between configurations of this form (coming from acyclic orientations with a unique sink) and maximal superstable configurations.

Theorem 8.3.3. *For a finite undirected graph G and for all divisors D on G,* **exactly one** *of the following holds:*

1. *D is equivalent to an effective divisor (i.e. the dollar game is winnable from D.)*

2. *There exists an acyclic orientation $\mathcal{O}$ of G such that the divisor*

$$(\text{outdeg}(\mathcal{O}) - \mathbf{1}) - D$$

is equivalent to an effective divisor.

Proof. Fix a vertex q of G to be the designated sink vertex. Suppose that D is not equivalent to an effective divisor. Then we construct an acyclic orientation of G as follows:

Let F be the unique q-reduced divisor (superstable configuration) linearly equivalent to D.

Let F' be a maximal superstable configuration such that $F' \geq F$ at all non-sink vertices. Declare $F'(q) = -1$ at the sink vertex.

Then F' is of the form $\text{outdeg}(\mathcal{O}) - \mathbf{1}$ for some acyclic orientation $\mathcal{O}$ of G by Theorem 3.5.1.

We claim that $F' - F$ is effective, i.e. $F' - F \geq 0$. This is true by construction for all vertices not equal to the sink. At the sink vertex, $F' = -1$ by definition. Also, at the sink vertex, F must be negative. Otherwise F would itself be effective, but it is linearly equivalent to D which we have assumed is not linearly equivalent to any effective divisor. Therefore, F has a negative value at the sink and $F' - F$ is effective.

The divisor D is linearly equivalent to F so $F' - D$ is linearly equivalent to $F' - F$ and in particular is linearly equivalent to an effective divisor. $\qquad \square$

Proof. (Of Theorem 8.3.1 Riemann–Roch for Graphs)

Let D be a divisor.

Claim:

$$r(D) - r(K - D) = \deg(D) + 1 - g.$$

Suppose $r(D) = \alpha$ and let $F \geq 0$ be a witness to the rank, i.e. suppose $\deg(F) = \alpha + 1$ and $D - F$ is not linearly equivalent to an effective divisor.

Since $D - F$ is not linearly equivalent to an effective divisor, by Theorem 8.3.3 there must be an acyclic orientation $\mathcal{O}$ of G such that $(\text{outdeg}(\mathcal{O}) - 1) - (D - F)$ is equivalent to an effective divisor. Let $E \geq 0$ be an effective divisor such that

$$(\text{outdeg}(\mathcal{O}) - 1) - (D - F) \sim E. \tag{8.1}$$

Let $\mathcal{O}'$ be the acyclic orientation obtained from $\mathcal{O}$ by reversing the orientation of all edges. Note that the sum of the divisors obtained from these two orientations is the canonical divisor:

$$(\text{outdeg}(\mathcal{O}) - 1) + (\text{outdeg}(\mathcal{O}') - 1) = K.$$

Adding $\text{outdeg}(\mathcal{O}') - 1$ to both sides of Equivalence 8.1 yields:

$$K - (D - F) \sim E + \text{outdeg}(\mathcal{O}') - 1$$

rearranging gives:

$$K - D - E \sim (\text{outdeg}(\mathcal{O}') - 1) - F.$$

This implies that $K - D - E$ is not linearly equivalent to an effective divisor because F is effective and so we can invoke Theorem 8.3.3. Therefore, by the definition of rank,

$$r(K - D) < \deg(E).$$

Now, Equation 8.1 implies

$$\deg(E) = \deg(\text{outdeg}(\mathcal{O}) - 1) - \deg(D) + \deg(F) \tag{8.2}$$
$$= (m - n) - \deg(D) + r(D) + 1. \tag{8.3}$$

Together,

$$r(K - D) < (m - n) - \deg(D) + r(D) + 1.$$

Rearranging gives,

$$r(D) - r(K - D) > \deg(D) - g, \tag{8.4}$$

which is very close to the Riemann–Roch claim. Role reversing K and $K - D$ yields Equation 8.5 which is complementary to Equation 8.3:

$$r(K - (K - D)) < (m - n) - \deg(K - D) + r(K - D) + 1. \quad (8.5)$$

Rearranging gives,

$$r(D) - r(K - D) < (m - n) - \deg(K - D) + 1.$$

Equivalently,

$$r(D) - r(K - D) < g - \deg(K - D). \quad (8.6)$$

Together, Equations 8.6 and 8.4 give:

$$\deg(D) - g < r(D) - r(K - D) < g - \deg(K - D)$$
$$\deg(D) - g < r(D) - r(K - D) < g - \deg(K) + \deg(D)$$
$$\deg(D) - g < r(D) - r(K - D) < g - (2g - 2) + \deg(D)$$
$$\deg(D) - g < r(D) - r(K - D) < \deg(D) - g + 2.$$

The bounded quantity $r(D) - r(K - D)$ is integer valued and hence equals $\deg(D) - g + 1$.

$\square$

Example 8.3.4. Let $G = K_4 \backslash e$ as shown below. The q-reduced divisors (superstable configurations) of G are listed in Example 2.6.20.

Consider the divisor:

$$D = v_3 - v_1 - v_2 - q.$$

The divisor D is not equivalent to an effective divisor. We are thus in case 2 of Theorem 8.3.3. Working through the steps of the proof of Theorem 8.3.3 we first determine the q-reduced divisor F equivalent to D. In this case,

$$F = v_1 - q.$$

Next we take a maximal q-reduced divisor F' such that $F' \geq F$. For example,

$$F' = v_1 + v_2 - q.$$

The divisor F' has the form

$$\text{outdeg}(\mathcal{O}) - 1$$

for an acyclic orientation $\mathcal{O}$ of G. Indeed the orientation orients vertices from smaller to larger subscript and from v_i to q. The vertex q is the unique sink of the orientation. The divisor $F' - D$ is linearly equivalent to $F' - F$ which can be checked explicitly using the Laplacian. The Laplacian appears (in reduced form) in Example 2.6.16.

Since D is not equivalent to an effective divisor, $r(D) = -1$. The Riemann–Roch theorem tells us about the rank of the error term $r(K - D)$.

$$K = v_2 + v_3$$
$$K - D = 2v_2 + v_1 + q.$$

$$(-1) - r(v_1 + 2v_2 + q) = -2 + 1 - 2$$
$$r(v_1 + 2v_2 + q) = 2.$$

The Riemann–Roch theorem has been extended and reinterpreted in a number of ways since the presentation by Baker and Norine. Backman [Bac17], for example, extends the result to directed graphs using the cycle–cocycle reversal systems of Section 4.7.3. Riemann–Roch theorems inspired by the Baker–Norine result also appear in [GK08], [MZ08], [AM10], [Lor12], [AC13], [JM13], [MS13], [AB15] and [CLB16].

Baker and Norine's original result itself is in fact more general than the result presented here. Their setup works over very general set systems equipped with an equivalence relation satisfying basic additivity properties. The general result is then applied to graphs and the firing equivalence relation.

The algebraic geometry narrative can be continued to other rank statements. For example, the following corollary is Clifford's Theorem for graphs.

Corollary 8.3.5 (Clifford's Theorem for graphs, [BN07]). *Let D be a divisor such that D and $K - D$ are both linearly equivalent to effective divisors. Then*

$$2r(D) \leq \deg(D).$$

Baker and Norine also define harmonic morphisms of graphs [BN09] leading, for example, to work on harmonic and other group actions on graphs; see e.g. [Cor10] and [GM14]. We refer the interested reader to Caporaso's survey on ranks of divisors [Cap13].

8.4 Torelli's theorem

In Chapter 4, we saw that the sandpile group, equivalently Jacobian or Picard group is an algebraic invariant of a graph. Furthermore, Wagner observed that for two graphs G_1 and G_2, if their graphical matroids are isomorphic then their sandpile groups are isomorphic,

$$M(G_1) \cong M(G_2) \implies \mathcal{S}(G_1) \cong \mathcal{S}(G_2),$$

and that the converse is false; see Section 4.4.

The fact that two non-isomorphic graphs can have isomorphic Jacobians is summarized by Bacher, de la Harpe and Nagnibeda as "a naive Torelli's Theorem does not hold" [BdlHN97].

Similar to the Bacher, de la Harpe, Nagnibeda constructions, which were in terms of lattices, Kotani and Sunada [KS00] define the Albanese and Jacobian *tori* of a graph and ask when two graphs have isometric tori.

Caporaso and Viviani [CV10] first answered this question, proving a Torelli's Theorem for graphs and tropical curves. Su and Wagner [SW10] prove an analogous result for all regular matroids.

First, we define the Albanese torus. As in Chapter 7, let $C_1(G; \mathbb{R})$ and $C_1(G; \mathbb{Z})$ denote the space of 1-chains of G with respectively real and integer coefficients. Define a scalar product on $C_1(G; \mathbb{R})$ by extending linearly from:

$$\langle e, f \rangle = \begin{cases} 1 & \text{if } e = f \\ 0 & \text{otherwise.} \end{cases}$$

Definition 8.4.1. For a finite graph G, the *Albanese torus* Alb(G) is

$$\text{Alb}(G) = (H_1(G; \mathbb{R})/H_1(G; \mathbb{Z}); \langle, \rangle)$$

with the flat metric induced by the scalar product.

Kotani and Sunada's *Jacobian torus* is the dual flat torus constructed via the cohomology groups: $\mathrm{Jac}(G) = \left(H^1(G; \mathbb{R}) / H^1(G; \mathbb{Z}); \langle, \rangle \right)$.

As above, for a graph G, let $M(G)$ denote the graphical matroid of G. Let $M_{\bullet}(G)$ denote the graphical matroid of G with all coloops contracted. Recall that a coloop of a graphical matroid is an edge that is contained in all spanning trees, i.e. an edge whose removal disconnects G.

Theorem 8.4.2 (Torelli's Theorem [CV10]). *For two graphs G_1 and G_2, their Albanese tori are isometric*

$$\mathrm{Alb}(G_1) \cong \mathrm{Alb}(G_2)$$

if and only if their contracted graphical matroids are isomorphic

$$M_{\bullet}(G_1) \cong M_{\bullet}(G_2).$$

Corollary 8.4.3 ([CV10]). *If G_1 and G_2 are 3-connected graphs then their Albanese tori are isometric*

$$\mathrm{Alb}(G_1) \cong \mathrm{Alb}(G_2)$$

if and only if the graphs are isomorphic

$$G_1 \cong G_2.$$

Corollary 8.4.3 follows from Theorem 8.4.2 and Whitney's Theorem. Whitney's Theorem characterizes when two graphs have the same graphical matroid; this holds if the two graphs can be obtained from each other through a sequence of graphical operations known as splittings, mergings and twistings. The theory also implies that for 3-connected graphs, the graphical matroid uniquely determines the graph.

The backwards direction of Torelli's Theorem, that isomorphic matroids imply isometric tori, is also not difficult equipped with Whitney's result. First, if the two graphs are connected, then they are connected by twistings alone. Second, if the contracted matroids of the two graphs are isomorphic then one graph can be obtained from the other in such a way that preserves cycles. From the cycles, explicit bases of the lattices $H_1(G_i; \mathbb{Z})$ can be constructed, see Exercise 4.8.12. The forward direction is considerably more involved, we refer the reader to [CV10].

8.5   The $\mathrm{Pic}^g(G)$ torus

The work of An, Baker, Kuperberg and Shokrieh [ABKS14] gives a combinatorial decomposition of the $\mathrm{Pic}^g(G)$ torus of a graph G with genus g, where $\mathrm{Pic}^g(G)$ is defined as $\mathrm{Div}^g(G)/\mathrm{Prin}(G)$ and identified with $H_1(G;\mathbb{R})/H_1(G;\mathbb{Z})$, here called the Jacobian torus $\mathrm{Jac}(G)$.

The decomposition is into parallelepipeds indexed by break divisors. Recall break divisors from Section 4.7.2 which were referred to as break configurations.

Definition 8.5.1. A *break divisor* of G is any divisor which consists of g chips total and which is formed by choosing a spanning tree T of G and placing, for each edge $e \notin T$, one chip at one of the endpoints of e.

Furthermore recall that for a graph of genus g, every divisor of degree g is linearly equivalent to a unique break divisor. We used this fact to construct the Bernardi sandpile torsor in Section 4.7.2. The uniqueness result originates in the work of Mikhalkin and Zharkov [MZ08] and a combinatorial proof is given in [ABKS14].

The decomposition result requires divisors on weighted graphs. We give only a brief idea of divisors in the weighted case.

A weighted graph is a finite graph equipped with a real valued weight associated to each edge. A weighted graph provides a *model* for a metric graph via the path metric along weighted edges. Divisors are supported at any finite collection of points on a metric graph, not only at the vertices. Chip-firing from an arbitrary point on a metric graph moves chips an epsilon distance in each direction.

For a graph G, each full dimensional cell in the ABKS decomposition of $\mathrm{Pic}^g(G)$ corresponds to a spanning tree of G. The interiors of the cells are parameterized by break divisors supported on the edges of G. Each vertex of the decomposition corresponds to a (integral) break divisor.

Figure 8.1 [ABKS14, Figure 1] shows the two-dimensional torus $\mathrm{Pic}^2(G)$ for the graph $G = K_4 \setminus e$ decomposed into parallelograms by break divisors. As in Exercise 4.8.12, the fundamental cycles can be used to generate an explicit basis for $H_1(G,\mathbb{Z})$. For this example, the genus of G, equal to the number of edges not in a spanning tree, is equal to two and hence the torus is two-dimensional. The eight two-dimensional cells correspond to the eight spanning trees of G. In each cell, the break divisor is supported on the collection of edges that forms the (edge) complement of a spanning tree.

An, Baker, Kuperberg and Shokrieh further use their decomposition of $\text{Pic}^g(G)$ to give a "geometric proof" of the Matrix-Tree Theorem; see [ABKS14].

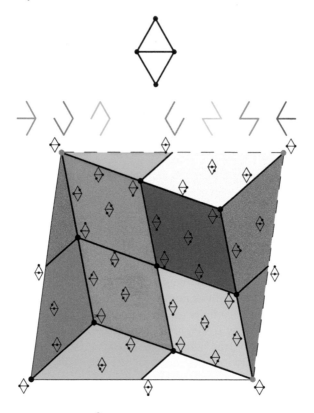

FIGURE 8.1: The Pic^2 torus for $K_4 \setminus e$ decomposed by break divisors [ABKS14, Figure 1].

8.6   Metric graphs and tropical geometry

Many of the chip-firing results inspired by the theory of curves can be further extended to metric graphs and into the domain of tropical geometry. The theories are not simply disjoint parallel stories. The classical and combinatorial settings have both influenced each other. We will not cover tropical geometry in this text, but point out a few notable results.

For example, after suitably defining the tropical Picard group, one has the following theorem.

Theorem 8.6.1 ([BF06],[HMY12]). *The $\mathbb{Q}$-tropical Picard group of a $\mathbb{Q}$-tropical curve Γ is the direct limit of the Picard groups corresponding to the subdivisions of Γ.*

Of particular importance in this area is Baker's specialization lemma [Bak08]. Without precise details, the specialization lemma states that the rank of divisor on an arithmetic surface is less than or equal to the rank of a corresponding divisor on a graph.

Lemma 8.6.2 ([Bak08]). *For a smooth curve X and all divisors $D \in \text{Div}(X)$*

$$r_G(\rho(D)) \geq r_X(D),$$

where ρ is the specialization map from a curve X to its dual graph G.

Tropical Riemann–Roch Theorems appear in [GK08], [MZ08] and [AC13]. A tropical Torelli's Theorem is presented along with the graphical case in [CV10]. The study of the complete linear system of a tropical curve appears in [HMY12]. And, in the other direction, a tropical proof of the Brill–Noether Theorem appears in [CDPR12].

Work has also been done to bring the tropical picture into higher dimensions; see for example, [Car13] which gives a higher-dimensional theory distinct from the one presented in Chapter 7.

8.7 Arithmetic geometry

Next we consider chip-firing from the motivation of arithmetic geometry. The theory of arithmetical graphs was introduced and developed by Lorenzini in a series of papers [Lor89, Lor91, Lor00, Lor12].

An arithmetical graph is a singular M-matrix with an associated distinguished integer vector which spans the kernel. Geometrically, arithmetical graphs are interpreted as degenerations of curves. The arithmetical information provides an intersection matrix and multiplicity vector for the graph.

Lorenzini further introduced a two-variable zeta-function associated to a graph motivated by the theory of two-variable zeta-functions for number fields. The zeta-function acts as a kind of generating function for divisors by rank and degree. Moreover, they are explicitly related to the Tutte polynomial of the graph via the rank generating function.

Before stating the two-variable zeta-function, we set up a general Riemann–Roch theory for lattices; for this we follow [Lor12].

Definition 8.7.1. For a directed multigraph G on n vertices and an integer vector $Q = (q_1, \ldots, q_n) \in \mathbb{Z}^n$, the *generalized Laplacian* $L(G,Q)$ is the $n \times n$ matrix with entries given by

$$L(G,Q)_{ij} = \begin{cases} -m_{i,j} & \text{if } i \neq j, \\ q_i & \text{if } i = j, \end{cases}$$

where $m_{i,j}$ is the number of directed edges from vertex i to vertex j.

Hence the generalized Laplacian matrix has the same off-diagonal entries as the graphical Laplacian for a directed multigraph. The diagonal however has been replaced by a fixed integer vector Q which is possibly different than the outdegree sequence of the graph.

Definition 8.7.2. An *arithmetical graph* is a triple (G, Q, R) consisting of a directed multigraph G and integer vectors $Q = (q_1, \ldots, q_n) > \mathbf{0}$ and $R = (r_1, \ldots, r_n) > \mathbf{0}$ such that:

1. $\gcd\{r_i\} = 1$ and

2. $L(G,Q)R^T = 0$.

The pair (Q, R) is called an *arithmetical structure* for G.

Lorenzini interpreted the Laplacians $L(G,Q)$ as the intersection matrices of degenerations of curves and R as the corresponding multiplicity vector.

Example 8.7.3. Any undirected graph G can be seen as an arithmetical graph as $(G, \deg(G), \mathbf{1})$. In this case, the generalized Laplacian reduces to the usual graphical Laplacian for G and, as we have already seen, the all ones vector is in the kernel of the graph Laplacian.

Example 8.7.4. Consider the matrix:

$$L(G,Q) = \begin{pmatrix} 3 & -1 & 0 & -1 \\ -1 & 2 & -1 & 0 \\ 0 & -1 & 3 & -1 \\ -1 & 0 & -1 & 1 \end{pmatrix}.$$

The off-diagonal entries reflect the adjacencies of the 4-cycle, C_4. The diagonal has been replaced with the primitive integer vector

$$Q = (3, 2, 3, 1).$$

The rank of $L(G, Q)$ is 3 which is $n - 1$. The vector

$$R = (1, 1, 1, 2)$$

is in the kernel of $L(G, Q)$:

$$L(G, Q)R^T = 0.$$

Therefore the pair (Q, R) is an arithmetical structure for C_4.

Notice that the matrix $L(G, Q)$ is a Z-matrix, where we recall from Chapter 6 that a Z-matrix is a matrix with non-negative diagonal entries and non-positive off-diagonal entries.

The matrices $L(G, Q)$ are in fact M-matrices. The important difference from Chapter 6 is that all M-matrices of Chapter 6 were assumed to be non-singular. In the graphical case, the *reduced* Laplacian was the relevant operator. The M-matrices of arithmetical graphs are necessarily singular matrices of rank $n - 1$.

Definition 8.7.5. The *sandpile group* $S(G, Q, R)$ of an arithmetical graph is the torsion (finite) part of the cokernel of $L(G, Q)$:

$$S(G, Q, R) = \ker(R^T)/\operatorname{im}(L(G, Q)).$$

Following the arithmetic geometry narrative, this group is called the *group of components*.

Definition 8.7.6. The *linear rank* $g_0(G)$ of an arithmetical graph (G, Q, R) is defined implicitly as

$$2g_0(G) - 2 = \sum_{i=1}^{n} r_i(q_i - 2).$$

The linear rank is the arithmetical equivalent to the genus g of the previous sections. In particular, the linear rank is always a positive integer.

Example 8.7.7. Let (G, Q, R) be the arithmetical graph of Example 8.7.4.

The linear rank is

$$2g_0(G) - 2 = 1(1) + 1(0) + 1(1) + 2(-1)$$
$$2g_0(G) = 2$$
$$g_0(G) = 1.$$

8.8   Riemann–Roch for lattices

Lorenzini extends the Riemann–Roch theory introduced by Baker and Norine to arithmetical graphs by defining a Riemann–Roch structure for lattices. Throughout, one should think of $R = 1$ for the graphical case.

Let $R = (r_1, \ldots, r_n) \in \mathbb{Z}^n$ be an integer vector with $\gcd\{r_i\} = 1$. For the remainder of the chapter, any integer vector $D \in \mathbb{Z}^n$ will be referred to as a *divisor*. For a divisor D define the *degree* of D with respect to R as:

$$\deg_R(D) = DR^T.$$

The degree map sends a divisor to its degree and the kernel of the degree map defines a lattice:

$$\Lambda_R = \{D \in \mathbb{Z}^n : DR^T = 0\}.$$

Let Λ be a rank $(n-1)$ sublattice of Λ_R. Define the *Picard group* of Λ as:

$$\mathrm{Pic}(\Lambda) = \mathbb{Z}^n / \Lambda,$$

and denote the class of a divisor D in $\mathrm{Pic}(\Lambda)$ as $[D]$.

$\mathrm{Pic}^0(\Lambda)$ is the kernel of the degree map from $\mathrm{Pic}(\Lambda)$ to $\mathbb{Z}$:

$$\mathrm{Pic}^0(\Lambda) = \Lambda_R / \Lambda.$$

Definition 8.8.1. Let Λ be a rank $(n-1)$ sublattice of Λ_R. The *g-number* of Λ is the smallest integer γ such that every divisor D of degree at least γ is equivalent to some effective divisor E; i.e. there exists $E > \mathbf{0}$ such that $D - E \in \Lambda$.

Clearly these definitions agree with those in Sections 8.1 and 8.2 in the graphical case.

Example 8.8.2. Let L be the usual graph Laplacian of a finite undirected graph G and let $R = \mathbf{1}$. The lattice Λ is equal to $\mathrm{im}(L)$ which is a rank $(n-1)$ sublattice of Λ_1. $\mathrm{Pic}^0(\Lambda)$ is the sandpile group of the graph. The g-number $g(\Lambda)$ is the genus of G, $g(\Lambda) = g(G) = m - n + 1$.

Example 8.8.3. Given an arithmetical graph (G, Q, R), let $\Lambda = \mathrm{im}\, L(G, Q)$. By construction, the lattice will have rank $(n-1)$. We expect that the g-number of the lattice will be equal to the linear rank. This is not generally the case, instead it provides an upper bound:

$$g(\Lambda) \leq g_0(L).$$

The g-number of a lattice is related to the Frobenius number of a lattice.

Definition 8.8.4. Given integers $a_1, \ldots, a_n \in \mathbb{Z}_{>0}$ such that $\gcd\{a_i\} = 1$, the *Frobenius number* is the largest integer that cannot be expressed as

$$a_1 x_1 + a_2 x_2 + \cdots + a_n x_n,$$

where the $x_i \geq 0$.

Determining the Frobenius number is also known as the *coin problem*. Suppose there are n different denominations of coins. The coin problem is to determine how large a quantity cannot be represented with the coins; see [BR15]. We mention that for fixed $n > 3$ finding the Frobenius number of a lattice is NP-Hard.

In our context, for a fixed $R = (r_1, \ldots, r_n) \in \mathbb{Z}_{>0}^n$, define $g(R) = g(r_1, \ldots, r_n)$ to be one more than the Frobenius number of the $\{r_i\}$.

Proposition 8.8.5. *Let $R = (r_1, \ldots, r_n)$, then*

$$g(\Lambda_R) = g(R) = g(r_1, \ldots, r_n).$$

Finally, in order to state a Riemann–Roch theory, we need a notion of canonical divisor.

Definition 8.8.6. Let $\Lambda \subseteq \Lambda_R$ be a lattice of rank $n - 1$ as above. A *canonical divisor* for Λ is a divisor K with $\deg(K) = 2g(\Lambda) - 2$ such that for all divisors D of degree $g(\Lambda) - 1$ either:

- both $[D]$ and $[K - D]$ contain an effective divisor or

- neither $[D]$ nor $[K - D]$ contains an effective divisor.

Not all lattices have canonical divisors. Even lattices coming from arithmetical graphs may not have canonical divisors. However we have the following:

Theorem 8.8.7. *For an arithmetical graph (G, Q, R), if $g(\Lambda) = g_0(L)$ then Λ has a canonical divisor.*

Lorenzini proved Theorem 8.8.7 using arithmetical geometry [Lor12]. Amani and Spencer provide a combinatorial proof of Theorem 8.8.7 in [AB11].

We can now define an abstract Riemann–Roch structure for a lattice.

Definition 8.8.8. For a positive integer vector $R \in \mathbb{Z}^n$ and a rank $(n-1)$ lattice $\Lambda \subseteq \Lambda_R$ with g-number g, a *Riemann–Roch structure* on Λ is a function

$$h : \mathrm{Pic}(\Lambda) \to \mathbb{Z}_{\geq 0}$$

such that:

1. There exists a divisor K such that for all divisors D,

$$h(D) - h(K - D) = \deg(D) + 1 - g.$$

2. If $[D] = [0]$ then $h([D]) = 1$,
   otherwise if $\deg(D) \leq 0$, then $h([D]) = 0$.

3. $h(D) \geq 1$ if and only if the class $[D]$ contains an effective divisor.

Again, not all lattices have Riemann–Roch structures. However we have the following:

Theorem 8.8.9 ([Lor12]). *A lattice $\Lambda \subseteq \Lambda_R$ of rank $n-1$ with g-number g and $|\mathrm{Pic}^0(\Gamma)| > 1$ has a Riemann–Roch structure if and only if there exists a canonical divisor K for Λ.*

Corollary 8.8.10. *For an arithmetical graph (G, Q, R), let $\Lambda = \mathrm{im}(L)$. If $g(\Lambda) = g_0(L)$ then there exists a Riemann–Roch structure for Λ.*

Example 8.8.11. For a finite undirected graph G, let $h(D) = r(D) + 1$, one more than the rank of the divisor. Then the conditions of Definition 8.8.8 are satisfied by the Riemann–Roch theory of Baker and Norine.

8.9   Two-variable zeta-functions

One of Lorenzini's primary motivations for defining general lattices with Riemann–Roch structures was to associate a zeta-function to the lattice. The two-variable zeta-function presented here is motivated by the two-variable zeta-function for number fields [VDGS00] and in turn by the two-variable zeta-function for a curve over a finite field [Pel96]; see [LR03] for more on this motivation from arithmetic geometry.

Definition 8.9.1. Let $\Lambda \subseteq \Lambda_r$ be a lattice of rank $n-1$ with g-number g and a Riemann–Roch structure h. The *zeta-function* of h is:

$$Z_h(\Lambda, t, u) = \sum_{[D] \in \mathrm{Pic}(\Lambda)} \frac{u^{h(D)} - 1}{u - 1} t^{\deg(D)}.$$

The zeta-function $Z_h(\Lambda, t, u)$ can be written as a rational function

$$Z_h(\Lambda, t, u) = \frac{f(t, u)}{(1 - t)(1 - tu)},$$

where $f(t, u)$ is a polynomial with a particularly nice form, which we will relate to the Tutte polynomial in Theorem 8.9.3.

We consider the zeta-function in relation to the following generating function for divisors.

Definition 8.9.2. Given a lattice $\Lambda \subseteq \Lambda_r$ of rank $n - 1$ with g-number g and Riemann–Roch structure h, define

$$W_h(\Lambda, x, y) = \sum_{[D] \in \mathrm{Pic}(\Lambda)} x^{h(D)} y^{h(K-D)}.$$

Formally, the functions W_h and Z_h are related as follows:

$$W_h(\Lambda, ut, t^{-1}) = (u - 1)t^{1-g} Z_h(\Lambda, t, u). \tag{8.1}$$

We consider the functions W_h and Z_h in the graphical case. Let G be a finite undirected graph and L the graph Laplacian for G. Let $\Lambda = \mathrm{im}\, L$ and h be equal to one more than the rank of a divisor. Then the two-variable zeta-function of G is

$$Z_G(t, u) = \sum_{[D] \in \mathrm{Pic}(G)} \frac{u^{r(D)+1} - 1}{u - 1} t^{\deg(D)}.$$

The function Z_G records divisor classes by rank and degree:

Z_{G_1} is equal to Z_{G_2} if and only if G_1 and G_2 have the same number of divisor classes of the same rank and degree.

Staying with the graphical case, the function W_h is reminiscent of the Tutte polynomial defined in terms of activity,

$$T(G, x, y) = \sum_{T \in \mathcal{T}(G)} x^{ia(T)} y^{ea(T)},$$

where $\mathcal{T}(G)$ is the collection of all spanning trees, $ea(T)$ is the external activity of a tree and $ia(T)$ is the internal activity of a tree; see Section 3.2.3.

Furthermore, recall Merino's Theorem which equated the specialization of the Tutte polynomial $T_G(1, y)$ to the critical polynomial; see Section 3.3. The critical polynomial is the generating function for the critical configurations of a graph by level. In the language of this chapter, the critical polynomial is the generating function for q-reduced divisors (which are dual to critical configurations) by a statistic that is essentially the degree (the level statistic adjusts the degree of a divisor based on the sink vertex.)

To emphasize the similarity to the zeta-polynomial, let $\mathcal{L}(G)$ be the generating function of q-reduced divisors by degree (superstable configurations by weight):

$$\mathcal{L}_G(t) = \sum_{q\text{-reduced divisors} D} t^{\deg(D)}.$$

Biggs considered $\mathcal{L}_G(t)$ as the growth function of the Picard group of a graph with respect to minimal presentations of the elements of the group [Big99b]. Biggs proved the identity:

$$T(G, 1, t^{-1}) = t^{-g}\mathcal{L}(G). \tag{8.2}$$

Equation 8.2 has the same form as Equation 8.1 with the Tutte polynomial comparing to the function W_h and the growth polynomial comparing to the zeta-polynomial.

The Tutte polynomial and the zeta-function are explicitly related in the next result.

Theorem 8.9.3 ([Lor12]). *For a connected graph G,*

$$\mathcal{L}(G) = f(t, 0).$$

Equivalently,

$$T(G, 1, t^{-1})t^g = Z_h(\Lambda, t, 0)(1 - t).$$

Example 8.9.4. The graph G of Figure 8.2 is a wedge of $K_4 \setminus e$ and C_3. The zeta-function is computed in [CLP15] and is equal to

$$Z_G(t, u) = 1 + 6t + 16t^2 + 6t^3 u + t^4 u^2 + \frac{24t^3}{(1 - t)(1 - tu)}.$$

The Tutte polynomial of G is equal to

$$T(G, x, y) = (x + x^2 + y)(x + 2x^2 + x^3 + y + 2xy + y^2).$$

From here we compute,

$$f(t,0) = 24t^3 + (1-t) + 6t(1-t) + 16t^2(1-t)$$
$$= 8t^3 + 10t^2 + 5t + 1,$$

and

$$T(G,1,t^{-1}) = (2 + \frac{1}{t})(4 + \frac{1}{t} + \frac{2}{t} + \frac{1}{t^2})$$
$$= 8 + \frac{10}{t} + \frac{5}{t^2} + \frac{1}{t^3}.$$

Thus

$$\mathcal{L}(G) = t^3 T(G,1,t^{-1}) = f(t,0).$$

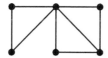

FIGURE 8.2: The graph of Example 8.9.4.

Lorenzini raised the question of when two graphs have the same zeta-function. As discussed earlier in Chapter 4, Clancy, Leake and Payne show that in considering the

- Tutte polynomial

- sandpile group and

- zeta function

of a graph, no two determine the third. Explicit constructions for each pair are provided in [CLP15].

8.10   Enumerating arithmetical structures

Given a graph G, little is known about the collection of distinct arithmetical structures that can be associated to G. What about the *number* of distinct arithmetical structures that can be associated to G?

Theorem 8.10.1 ([Lor89]). *There exist only finitely many arithmetical structures on any fixed graph.*

The proof of Theorem 8.10.1 proceeds in two steps. First, the following property of generalized Laplacians is established: if M is a generalized Laplacian of an arithmetical graph, then $M + X$ is non-singular for all non-negative, non-zero diagonal matrices X. Second, one recognizes that there are no infinite antichains in the componentwise partial order on integer sequences of a fixed length.

For special classes of graphs, the number of arithmetical structures admit combinatorial formulas. Let $\mathcal{A}(G)$ be the set of arithmetical structures for a graph G.

Theorem 8.10.2 ([CV18], [BCC$^+$17]).

- *The number of arithmetical structures on the path graph of length n, P_n, is the $(n-1)$st Catalan number,*

$$|\mathcal{A}(P_n)| = \frac{1}{n}\binom{2n-2}{n-1}.$$

- *The number of arithmetical structures on the cycle graph of length n, C_n, is $2n-1$ times the $(n-1)$st Catalan number,*

$$|\mathcal{A}(C_n)| = \binom{2n-1}{n-1}.$$

These enumerations are refined in [BCC$^+$17], the number of arithmetical structures is related to other combinatorial sequences such as the ballot sequences.

The number of arithmetical structures for complete graphs is unknown; it is conjectured however that they have the maximal number possible.

Conjecture 8.10.3 ([CV18]). *For any graph G on n vertices,*

$$|\mathcal{A}(P_n)| \leq |\mathcal{A}(G)| \leq |\mathcal{A}(K_n)|.$$

8.11 Exercises

Exercise 8.11.1. *Prove the claim after Definition 8.2.2:*
ρ_s *is harmonic, maps to 0 and the image generates all of* $\mathrm{Pic}(G)$.

Exercise 8.11.2. *Compute the rank of the divisors below.*

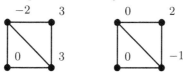

Exercise 8.11.3. *Prove that if* $\deg(D) \geq g$ *then the dollar game is winnable; i.e.* D *is linearly equivalent to an effective divisor.*

Exercise 8.11.4. *Prove that if the dollar game is winnable, then it is winnable via a greedy strategy: If at anytime any site is in debt, have the site borrow from its neighbors as much as possible.*

Exercise 8.11.5. *Let* G *be a graph with sink vertex* q. *Let* D *be a divisor on* G *and* S *the unique* q-*reduced divisor linearly equivalent to* D. *Prove that* D *is linearly equivalent to an effective divisor if and only if the value of* S *at* q *is non-negative.*

Exercise 8.11.6. *The Pentagon Problem.*

Let G *be a 5-cycle (a pentagon). Let* **c** *be an initial configuration on* G *allowing for both positive and negative values but such that the sum of all values over all vertices is positive. If at any time there is a vertex* v *with a negative value* $\mathbf{c}_v$ *then the value at* v *is added to both neighbors of* v *and the value at* v *is negated. Prove that this process terminates, showing in particular, that all vertices can be given a non-negative value through this process.*

Exercise 8.11.7. *Let* D *and* F *be two divisors with non-negative rank.*

- *Prove that* $r(D + F) \geq r(D) + r(F)$.

- *Prove Clifford's Theorem.*

Exercise 8.11.8. *Define the* gonality *of a graph* G *to be the minimum value* d *for which there exists a divisor of degree* d *on* G *with rank at least 1.*

Prove that the complete graph K_n *has gonality* $n - 1$.

Exercise 8.11.9. *Prove that the generalized Laplacians of arithmetical graphs are (singular)* M-*matrices by proving that all principal minors of a generalized Laplacian have positive determinant.*

Exercise 8.11.10. *Let* $G = K_4 \setminus e$. *Find all arithmetic structures for* G.

Exercise 8.11.11. *Show that the linear rank is equal to the genus for an arithmetical graph when M is the usual graph Laplacian.*

Exercise 8.11.12. *Prove that the linear rank can be given in terms of the degree sequence:*

$$2g_0(G) - 2 = \sum_{i=1}^{n} r_i(\deg(v_i) - 2).$$

Exercise 8.11.13. *Prove that the linear rank is always a positive integer.*

Exercise 8.11.14. *Compute the zeta-polynomial and Tutte polynomial for the graph that is a wedge of C_3 and $K_4 \setminus e$ with the wedge point at a vertex of degree two in $K_4 \setminus e$.*

Exercise 8.11.15. *Following the proof sketch in the text, prove that for a fixed graph G there are only finitely many arithmetic structures (G, Q, R).*

Chapter 9

Ideals

In this chapter we consider chip-firing from the perspective of combinatorial commutative algebra. The commutative algebra approach was first initiated by Cori, Rossin, and Salvy [CRS02]. Cori, Rossin, and Salvy were looking for efficient methods to compute aspects of the sandpile model. For example, they sought to compute the identity element of the Picard group by finding Gröbner bases with a small number of elements.

The two main objects of study are a binomial lattice ideal, the *toppling ideal*, and a corresponding monomial initial ideal, the *tree ideal*. The tree ideal was further studied by Postnikov and Shapiro [PS04] in relation to G-parking functions. The work of Postnikov and Shapiro, and later works by Dochtermann and Sanyal [DS14], Mohammadi and Shokrieh [MS16], and Manjunath, Schreyer and Wilmes [MSW15] all consider resolutions of these ideals. In particular, they give a combinatorial formula for the Betti numbers of the tree and toppling ideals as first conjectured in Perkinson, Perlman, and Wilmes [PPW13]. Manjunath and Sturmfels [MS13] further developed the commutative algebra perspective into a Riemann–Roch theory for monomial ideals using Alexander duality.

9.1 Toppling ideals

Chip configurations will now be formalized as Laurent monomials.

Let G be a finite graph with n vertices and $\mathbf{c}$ a chip-configuration with c_i chips at vertex i. Define the *chip configuration monomial* $\mathbf{x}^{\mathbf{c}}$ as:

$$\mathbf{x}^{\mathbf{c}} = x_1^{c_1} x_2^{c_2} \cdots x_n^{c_n}.$$

Let Δ be the graph Laplacian associated to G and let $\Lambda(\Delta)$ be the lattice generated by the columns of Δ,

$$\Lambda(\Delta) = \mathrm{im}_{\mathbb{Z}} \Delta.$$

Chip-firing moves are represented as binomials of the form:

$$\mathbf{x^c} - \mathbf{x^d},$$

where a *legal* binomial is one in which $\mathbf{c}, \mathbf{d} \geq \mathbf{0}$ and

$$\mathbf{c} - \mathbf{d} \in \Lambda(\Delta).$$

Namely, a legal binomial has exponents $\mathbf{c}$ and $\mathbf{d}$ which are non-negative elements of the same chip-firing equivalence class.

Definition 9.1.1. The *toppling ideal* of a graph G with n vertices and Laplacian Δ is the binomial ideal generated by all legal chip-firing binomials:

$$I_G = \langle \mathbf{x^u} - \mathbf{x^v} \mid \mathbf{u} - \mathbf{v} \in \Lambda(\Delta) \rangle$$

in the polynomial ring $k[x_1, x_2, \ldots, x_n]$.

The toppling ideal is an example of a lattice ideal which is a generalization of toric ideals. We refer to [MS05] for the combinatorial theory of ideals.

We can construct an explicit generating set for this ideal as follows. For each column of Δ, construct a binomial of the form $\mathbf{x^u} - \mathbf{x^v}$ where $\mathbf{u}$ corresponds to the positive entries in the column and $\mathbf{v}$ corresponds to the negative entries in the column. Specifically, define:

$$f_j^+(\mathbf{x}) = \prod_{k : \Delta_{kj} > 0} x_k^{\Delta_{kj}},$$

$$f_j^-(\mathbf{x}) = \prod_{k : \Delta_{kj} < 0} x_k^{-\Delta_{kj}},$$

$$f_j(\mathbf{x}) = f_j^+(\mathbf{x}) - f_j^-(\mathbf{x}).$$

Let $I_0 = \langle f_1(\mathbf{x}), \ldots, f_q(\mathbf{x}) \rangle$. The toppling ideal of G is then the saturation of I_0 with respect to the product of all the variables

$$I_G = I_0 : \langle x_1 \cdots x_n \rangle.$$

Example 9.1.2. Let G be $K_4 \setminus e$ as seen below.

The Laplacian $\Delta(G)$ is

$$\Delta(K_4 \backslash e) = \begin{array}{c} \\ v_1 \\ v_2 \\ v_3 \\ q \end{array} \begin{array}{cccc} v_1 & v_2 & v_3 & q \\ \left(\begin{array}{cccc} 2 & -1 & -1 & 0 \\ -1 & 3 & -1 & -1 \\ -1 & -1 & 3 & -1 \\ 0 & -1 & -1 & 2 \end{array}\right). \end{array}$$

Letting x_4 correspond to q, the toppling ideal of G is

$$I_G = \langle x_1^2 - x_2 x_3, x_2^3 - x_1 x_2 x_4, x_3^3 - x_1 x_2 x_4, x_4^2 - x_2 x_3,$$
$$x_1 x_3^2 - x_2^2 x_4, x_1 x_2^2 - x_3^2 x_4 \rangle.$$

The ideal I_0 is generated by the first four listed terms of I_G.

The ideal I_G is sometimes referred to as the *homogeneous toppling ideal*. Cori, Rossin, and Salvy [CRS02] originally introduced an inhomogeneous version. The homogeneous version presented here was introduced in [PPW13].

9.2 Tree ideals

In this section we introduce a monomial initial ideal M_G for the binomial toppling ideal I_G. While the binomial toppling ideal reflects firing equivalence, the monomial *tree ideal* reflects stabilization: the standard monomials of M_G are in bijection with the q-reduced divisors (superstable configurations) of G. Hence the number of standard monomials is precisely the number of spanning trees of G.

In order to construct an initial ideal of I_G, we must first specify a term order on chip-firing monomials. The term ordering should respect chip-firing dynamics: A legal chip-firing move should replace a monomial with a smaller monomial in the term order (a Gröbner reduction step). Generally, for two chip configurations $\mathbf{c}$ and $\mathbf{d}$, if $\mathbf{d}$ can be obtained from

c by a legal sequence of firings, then we want that $\mathbf{x^c} > \mathbf{x^d}$ with respect to the term ordering on monomials. We will use a rooted spanning tree to induce such a term order.

Let G be a finite graph on n vertices with a designated sink vertex q. Let T be a spanning tree of G. From each vertex v, there is a unique shortest path from v to the sink q. Define a partial order on the vertices of G as:

$$v_i \succeq_T v_j$$

if v_j lies on the unique shortest path from v_i to the sink. Fix a total ordering $\geq_T$ on the variables $x_1, x_2, \ldots, x_n$ by taking any linear extension of $\succeq_T$.

Definition 9.2.1. A *spanning tree monomial term order* is the graded reverse lexicographical (grevlex) term order with respect to any linear extension $\geq_T$ for a spanning tree T.

Proposition 9.2.2. *For any spanning tree monomial term order, if the configuration* **d** *can be obtained from the configuration* **c** *through a sequence of legal firings, then* $x^{\mathbf{c}} > x^{\mathbf{d}}$.

Definition 9.2.3. For a graph G with a sink q, the *tree ideal* M_G is the monomial initial ideal of the toppling ideal I_G with respect to any spanning tree monomial order formed by a tree rooted at q.

Implicit in the definition is that for a graph G with a fixed sink q, any spanning tree monomial term order with respect to q gives the same initial ideal of I_G. The ideal M_G may be different for different choices of q, just as the collection of superstable configurations can change for different choices of sink vertex.

Recall that monomial ideals may be thought of as complements of staircases in $\mathbb{Z}_{\geq 0}^n$; see Figure 9.1. This means that the collection of monomials in an ideal is upward closed in the componentwise partial order. For a monomial J, if

$$x_1^{\alpha_1} x_2^{\alpha_2} \cdots x_n^{\alpha_n} \in J$$

and $x_1^{\beta_1} x_2^{\beta_2} \cdots x_n^{\beta_n}$ is such that $\alpha_i \leq \beta_i$ for all i then

$$x_1^{\beta_1} x_2^{\beta_2} \cdots x_n^{\beta_n} \in J.$$

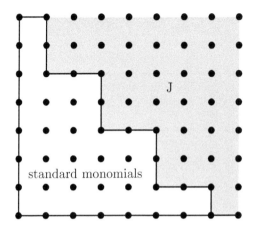

FIGURE 9.1: For a monomial ideal J, the (exponents of the) monomials not in J form a downward closed subset of $\mathbb{Z}_{\geq 0}^n$.

The *standard monomials* of a monomial ideal $J \subset R$, $R = k[x_1, x_2, \ldots x_n]$ are all those monomials not in J. The standard monomials form a basis for the algebra R/J. The ideal J is upward closed and hence the standard monomials are downward closed in the componentwise partial order.

Theorem 9.2.4. *The exponent vectors of the standard monomials of M_G are precisely the q-reduced divisors (superstable configurations) of G. Therefore the number of standard monomials of M_G is equal to the number of spanning trees of G.*

The staircase structure of standard monomials is consistent with our earlier study of superstable configurations. The fact that superstable configurations form a downward closed set in the componentwise partial order was established in Proposition 2.6.21.

Postnikov and Shapiro investigate M_G in the context of parking functions which, as we have seen, are closely related to q-reduced divisors; see Section 6.5.1. In particular, M_G can be described in terms of outdegrees of vertices of G. Let G be a graph on n vertices and let $I \subseteq V$ be any non-empty subset of vertices of G. Define a monomial

$$m_I = \prod_{i \in I} x_i^{\mathrm{outdeg}_I(i)},$$

where $\mathrm{outdeg}_I(i)$ is the outdegree of i with respect to I, the number of edges from i to vertices not in I.

Proposition 9.2.5. *The tree ideal M_G is generated by the monomials m_I where I runs over all non-empty subsets of non-sink vertices,*

$$M_G = \langle m_I : I \subseteq V \setminus q \rangle.$$

One can further describe the minimal generators of M_G combinatorially. For $I \subseteq [n]$, let $G[I]$ be the induced subgraph of G on I.

Proposition 9.2.6. *For a graph G, the monomial m_I is a minimal generator of M_G if and only if both $G[I]$ and $G[I^c]$ are connected.*

Example 9.2.7. Let $G = K_4 \setminus e$ as in Example 9.1.2 and shown below.

Let T be the spanning tree consisting of edges $\{v_1v_2, v_2q, v_3q\}$ so that $v_1 >_T v_2 >_T v_3 >_T q$ is a linear extension of the partial order induced by T rooted at q.

The tree ideal M_G is generated by the monomials m_I of outdegrees running over all subsets $I \subseteq \{1, 2, 3\}$:

$$M_G = \langle x_1^2, x_2^3, x_3^3, x_1x_2^2, x_1x_3^2, x_2^2x_3^2, x_2x_3 \rangle.$$

For $I = \{2, 3\}$, $G[I^c]$ is not connected:

In this case, $m_I = x_2^2x_3^2$, and indeed, the monomial $x_2^2x_3^2$ is not minimal in M_G.

The standard monomials of M_G are

$$\{1, x_1, x_2, x_2^2, x_3, x_3^2, x_1x_2, x_1x_3\}.$$

The exponent vectors of the standard monomials of M_G are

$$\{000, 100, 010, 020, 001, 002, 110, 101\}.$$

We confirm these are the eight q-reduced divisors of G by comparison to Example 2.6.20.

9.3   Resolutions

Free resolutions of the tree ideal appear in the works [PS04], [MS13] and [PPW13]. In a convergence of papers, minimal resolutions are constructed for M_G in [DS14], [MS16], and [MSW15]. We follow the description of [DS14] which provides a minimal *cellular* resolution of M_G. We recommend [MS05] for a comprehensive treatment of the combinatorial commutative algebra used here.

9.3.1   Cellular resolutions

To construct a free resolution of the tree ideal, we form an exact sequence

$$0 \leftarrow M \xleftarrow{\phi_0} F_1 \xleftarrow{\phi_1} \cdots \xleftarrow{\phi_r} F_r \leftarrow 0,$$

of graded R-modules

$$F_i \cong \bigoplus_{\mathbf{a} \in \mathbb{Z}^n} R(-\mathbf{a})^{\beta_i, \mathbf{a}},$$

with the $\mathbb{Z}^n$-grading and degree preserving maps ϕ_i.

Cellular resolutions use labeled cell complexes to construct resolutions of ideals. The exact sequence above is given by the chain complex of a cell complex with the graded maps given by boundary maps. The cell complex is labeled with each face labeled by a monomial term. This cellular free complex gives a resolution of the ideal generated by the monomial labels of vertices if the cell complex satisfies certain acyclic conditions. Thus the algebraic conditions of a resolution can be checked directly on the labeled cell complex.

For the tree ideal, the cell complex used to give a resolution arises from a hyperplane arrangement. Recall from Chapter 3 that for a finite simple graph G on n vertices, the *graphical* arrangement $\mathcal{A}_G \in \mathbb{R}^n$ is a sub-arrangement of the Braid arrangement consisting of the hyperplanes:

$$\{x_i = x_i\}, \ ij \in E(G).$$

Suppose G is a graph with n non-sink vertices and sink vertex $n+1$. Define the affine subspace U as follows:

$$U = \{x \in \mathbb{R}^{n+1} \mid x_{n+1} = 0, \ x_1 + x_2 + \cdots + x_n = 1\}.$$

Define $\tilde{\mathcal{A}}_G$ to be the affine hyperplane arrangement formed by intersecting $\mathcal{A}_G$ with U. Finally, let $\mathcal{B}_G$ be the bounded complex (complex of bounded regions) of the arrangement $\tilde{\mathcal{A}}_G$ in the ambient space U; see Figure 9.2.

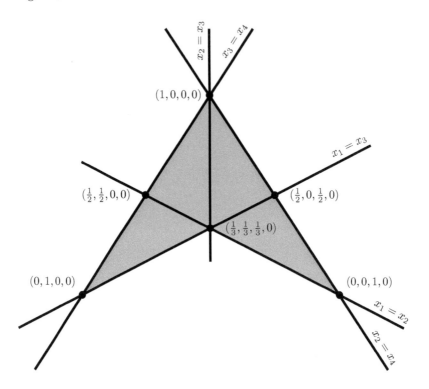

FIGURE 9.2: The bounded region for $K_4 \setminus e$.

The complex $\mathcal{B}_G$ supports a cellular resolution of the tree ideal M_G. In order to see this, we describe a labeling of the cells of $\mathcal{B}_G$.

Let $\mathbf{v} \in \mathbb{R}^{n+1}$ be a vertex of $\mathcal{B}_G$ and let $I_{\mathbf{v}} = \operatorname{supp}(\mathbf{v})$ be the support of $\mathbf{v}$, i.e. the set of indices corresponding to non-zero entries of $\mathbf{v}$. Note that since $\mathcal{B}_G$ lives in U, $n + 1$ will not be in the support $I_{\mathbf{v}}$ for any vertex $\mathbf{v}$ of $\mathcal{B}_G$.

Label the complex $\mathcal{B}_G$ as follows:

- For a vertex $\mathbf{v} \in \mathcal{B}_G$, label $\mathbf{v}$ with the monomial $m_{I_{\mathbf{v}}}$.

- For a face $\mathbf{f} \in \mathcal{B}_G$ of higher dimension, label $\mathbf{f}$ by the lcm of all labels of all vertices of $\mathbf{f}$.

Theorem 9.3.1 ([DS14]). *The complex $\mathcal{B}_G$ with the labeling above gives a minimal cellular resolution of M_G.*

Example 9.3.2. Let G be $K_4 \setminus e$ as in Example 9.1.2.

The graphical arrangement $\mathcal{A}_G$ intersected with the affine subspace U is depicted in Figure 9.2. The shaded region is the bounded region $\mathcal{B}_G$.

Figure 9.3 shows the bounded region $\mathcal{B}_G$ with the monomial labels on all vertices and 2-cells.

The f-vector of $\mathcal{B}(G)$ is

$$f(\mathcal{B}_G) = (1, 6, 9, 4).$$

The resolution of M_G is thus of the form

$$0 \leftarrow R \leftarrow R^6 \leftarrow R^9 \leftarrow R^4 \leftarrow 0,$$

with boundary maps given by the incidence structure of $\mathcal{B}(G)$.

9.3.2 Betti numbers

One corollary of Theorem 9.3.1, appearing in [DS14], is a proof of a combinatorial formula for the Betti numbers of M_G which was first conjectured by Wilmes [PPW13].

The combinatorial formula also appears in [Hop14], [MSW15] and [MS16]. In the last two works, it is further shown that the Betti numbers of M_G coincide with the Betti numbers of I_G. In order to describe these Betti numbers, consider the following definition.

Let Π be the collection of all partitions of the vertices of G such that G restricted to each part of the partition is connected. Let Π_k denote the partitions in Π with exactly k parts. For a partition $\pi \in \Pi$ of G, let G_π denote the graph formed by contracting all edges within each part of π.

Theorem 9.3.3 (Wilmes' Theorem). *For a finite connected graph G,*

$$\beta_k(R/M_G) = \beta_k(R/I_G)$$
$$= \sum_{\pi \in \Pi_{k+1}} |\{\text{maximal } q\text{-reduced divisors of } G_\pi\}|.$$

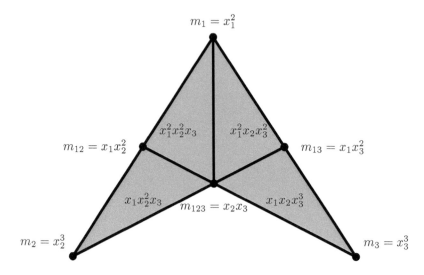

FIGURE 9.3: The bounded region for $K_4 \setminus e$.

In various sources the result is phrased in terms of certain G-parking functions, minimal recurrent configurations or acyclic orientations with a unique source, all of which are equinumerous; see Section 3.5.

In [MS16], the connected partitions of a graph G are considered with respect to the graphical matroid of G. The toppling and tree ideals are related to the graphic Lawrence ideal and graphic oriented matroid ideal. Moreover, cellular resolutions are provided for both the toppling and tree ideal.

Further extensions of the results in this section can be found in [MS14], [Moh16], [Doc17] and [Doc18].

9.4   Critical ideals

Before turning to the ideal theoretic Riemann–Roch Theorem, we briefly mention another ideal structure related to the chip-firing process. Toppling ideals are formed from the columns of the Laplacian. The critical ideal introduced by Corrales and Valencia [CV13], [AV14], [CV15], is a determinantal ideal formed from the Laplacian.

Let G be a finite directed graph and associate indeterminants

$$X_G = \{x_v \mid v \in V(G)\}$$

to each vertex of G. The generalized Laplacian of G is

$$L(G, X_G)_{ij} = \begin{cases} -m_{i,j} & \text{if } i \neq j, \\ x_i & \text{if } i = j, \end{cases}$$

where $m_{i,j}$ is the number of edges from vertex i to vertex j.

Definition 9.4.1. The kth *critical ideal* of G is the determinantal ideal generated by all $k \times k$ minors of L.

$$I_k(G, X_G) = \langle \det L(G, X_G)_{[k,k]} \rangle.$$

For $k \leq 0$, we define $I_k = \langle 1 \rangle$ and for $k > n$, we define $I_k = \langle 0 \rangle$.

Example 9.4.2. Let G be $K_4 \setminus e$ as in Example 9.1.2.

Associate x_i to the vertex v_i and let x_4 correspond to q, then the generalized Laplacian of G is

$$L(K_4 \setminus e, X_{K_4 \setminus e}) = \begin{array}{c} \\ v_1 \\ v_2 \\ v_3 \\ q \end{array} \begin{array}{c} v_1 \quad v_2 \quad v_3 \quad q \\ \begin{pmatrix} x_1 & -1 & -1 & 0 \\ -1 & x_2 & -1 & -1 \\ -1 & -1 & x_3 & -1 \\ 0 & -1 & -1 & x_4 \end{pmatrix} \end{array}.$$

For $k \leq 2$, the critical ideal is $I_k = \langle 1 \rangle$.
For $k = 4$, the critical ideal is generated by 1 term, the determinant of the generalized Laplacian Δ.
For $k = 3$, the critical ideal is generated by 10 terms:

$$I_3 = \langle -x_2 - x_3 - 2, x_1 x_3 + x_1, -x_1 x_2 - x_1, x_1 x_2 x_3 - x_1 - x_2 - x_3 - 2,$$
$$x_4 x_2 + x_4, -x_1 x_4 - x_4 - x_1, x_4 x_1 x_2 - x_4 - x_1, -x_4 x_3 - x_4,$$
$$x_4 x_2 x_3 - x_4 - x_2 - x_3 - 2, x_4 x_1 x_3 - x_4 - x_1 \rangle.$$

A Gröbner basis for I_3 is $\{x_2 x_3 x_4 + 2x_1\}$.

Recall from Chapter 4 that the sandpile group has gone by many names including the critical group. The name critical ideal reflects the close relationship to the critical group.

Proposition 9.4.3. *Let G be a graph with critical group $K(G)$ and invariant factors $f_1, f_2, \ldots, f_m$. Let $I_k(G)$ be the critical ideal taken with respect to the usual graph Laplacian, then $I_k(G)$ is generated by the product of the first k invariant factors of $K(G)$:*

$$I_k(G) = \langle \prod_{i=1}^{k} f_i \rangle.$$

The critical ideal can also be seen as a generalization of the characteristic polynomials of the adjacency and Laplacian matrices; see [CV13].

Definition 9.4.4. The *algebraic corank* of a graph G, $\gamma(G)$, is defined as

$$\gamma(G) = \max\{k \mid I_k(G, X_G) = \langle 1 \rangle\}.$$

The algebraic corank is related to well-studied combinatorial parameters of graphs including the independence and clique numbers. Furthermore, the algebraic corank distinguishes extremal families of graphs:

$$\gamma = 0 \quad \text{iff} \quad G = K_1$$
$$\gamma = 1 \quad \text{iff} \quad G = K_n \ n > 1$$
$$\gamma = n - 1 \quad \text{iff} \quad G = P_n.$$

The value $\gamma = n - 1$ is the maximum possible value for γ for a graph with n vertices.

Example 9.4.5. For our example above, $G = K_4 \setminus e$, the algebraic corank is

$$\gamma(K_4 \setminus e) = 2.$$

9.5   Riemann–Roch for monomial ideals

A Riemann–Roch theory for monomial ideals was developed by Manjunath and Sturmfels in [MS13]. As in the arithmetic geometry case, the Riemann–Roch theory is set up for general monomial ideals and then shown to hold for sufficiently nice ideals. First, we must define the notions of rank, degree, and canonical divisor in this setting. We require a number of technical definitions; these are illustrated in Example 9.5.6.

Throughout this section, let $M \subset k[x_1, \ldots, x_n]$ be an artinian monomial ideal, i.e. let M have only finitely many standard monomials.

Definition 9.5.1. The *rank* of a monomial $\mathbf{x^b}$ with $\mathbf{b} \geq \mathbf{0}$ is given by:

$$r(\mathbf{x^b}) = \min\{\deg(\mathbf{x^a}) \mid \mathbf{x^b} \in \langle \mathbf{x^a} \rangle \setminus \mathbf{x^a} M\} - 1.$$

Notice the minus one at the end of the definition. If the monomial $\mathbf{x^b} \in M$ then $r(\mathbf{x^b}) \geq 0$. If $\mathbf{x^b} \notin M$ then $r(\mathbf{x^b}) = -1$.

Definition 9.5.2. A *socle monomial* of M is a monomial $\mathbf{x^b}$ such that $\mathbf{x^b} \notin M$ but $\mathbf{x^b} x_i \in M$ for all x_i.

Informally, in our staircase imagery, socle elements sit in the corners of the steps. Let $\mathrm{MonSoc}(M)$ denote the collection of all socle elements of M. A monomial ideal M is *reflection invariant* if there exists a distinguished monomial $\mathbf{x^K}$ such that the map

$$\Phi : \mathbf{x^c} \to \mathbf{x^K}/\mathbf{x^c}$$

is an involution on $\mathrm{MonSoc}(M)$.

Definition 9.5.3. For a reflection invariant monomial ideal M, the monomial $\mathbf{x^K}$ is called the *canonical monomial* of M.

Proposition 9.5.4. *An artinian monomial ideal is reflection invariant with canonical monomial* $\mathbf{x^K}$ *if and only if the Alexander dual of M with respect to* $\mathbf{K} + \mathbf{1}$ *is the ideal generated by the socle elements of M:*

$$M^{[\mathbf{K}+\mathbf{1}]} = \langle \mathrm{MonSoc}(M) \rangle.$$

A monomial ideal M is *level* if all socle elements of M have the same degree.

Definition 9.5.5. For a level ideal M, the *genus* of M, $g(M)$ is equal to one more than the degree of any socle element.

Example 9.5.6. Consider the monomial ideal

$$M = \langle x^8, x^6 y^2, x^4 y^4, x^2 y^6, y^8 \rangle.$$

The socle elements are

$$\mathrm{MonSoc}(M) = \{x^7 y, x^5 y^3, x^3 y^5, xy^7\}.$$

x^8y^8

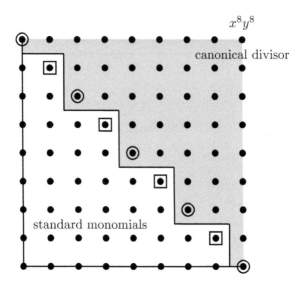

FIGURE 9.4: The generators of M are marked with circles. The elements of $\operatorname{MonSoc}(M)$ are marked with squares.

The generators of M and the elements of $\operatorname{MonSoc}(M)$ are shown in Figure 9.4. Figure 9.4 is based on Figure 2 of [MS13].

The degree of all socle elements is equal to 8; hence the ideal M is level. M is also reflection invariant with canonical monomial $\mathbf{x}^{\mathbf{K}} = x^8y^8$.

Informally, the rank of an element $\mathbf{z}$ of M corresponds to the minimum number of rectilinear shifts (minus one) needed to capture the element $\mathbf{z}$ by the solid staircase path bounding the standard monomials of M. The standard monomials themselves have rank -1.

We can now state the Riemann–Roch Theorem for monomial ideals.

Theorem 9.5.7 (Riemann–Roch for monomial ideals [MS13, Theorem 13]). *Let M be an artinian, level, reflection-invariant monomial ideal. Let $x^{\mathbf{K}}$ be the canonical divisor for M. Then for a monomial $x^{\mathbf{b}}$,*

$$r(x^{\mathbf{b}}) - r(x^{\mathbf{K}}/x^{\mathbf{b}}) = \deg(x^{\mathbf{b}}) - g(M) + 1.$$

As in the case for graphs, we also have a Clifford's Theorem.

Corollary 9.5.8 (Clifford's Theorem for monomial ideals [MS13]). *Let M be an artinian, level, reflection-invariant monomial ideal. Let $x^{\mathbf{b}}$*

be a monomial that divides $x^{\mathbf{K}}$. Further assume that $r(x^{\mathbf{b}}) \geq 0$ and $r(x^{\mathbf{K}}/x^{\mathbf{b}}) \geq 0$. Then

$$r(x^{\mathbf{b}}) \leq (\deg(x^{\mathbf{b}}) - 1)/2.$$

For *saturated* graphs, i.e. graphs with at least one edge between every pair of vertices, the tree ideal M_G satisfies the conditions of Theorem 9.5.7 and the Riemann–Roch relation holds. For non-saturated graphs, M_G may not be reflection invariant. Despite this, Manjunath and Sturmfels show how to derive the Baker–Norine Riemann–Roch Theorem for graphs from the Riemann–Roch Theorem for monomial ideals, see [MS13].

9.6 Exercises

Exercise 9.6.1. *Let $G = C_4$ be the four cycle.*

1. *Determine the toppling ideal I_G of G.*

2. *Determine the tree ideal M_G of G.*

3. *Confirm that the standard monomials of M_G correspond to the superstable configurations of G.*

Exercise 9.6.2. *Prove that for any spanning tree monomial term order, if the configuration $\mathbf{d}$ can be obtained from the configuration $\mathbf{c}$ through a sequence of legal firings, then $x^{\mathbf{c}} > x^{\mathbf{d}}$.*

Exercise 9.6.3. *Prove that for a graph G with a fixed sink q, any spanning tree monomial term order with respect to q gives the same initial ideal of I_G.*

Exercise 9.6.4. *Let $G = K_4 \setminus e$.*

1. *Compute the set Π for G.*

2. *Partially order Π under refinement.*

3. *Confirm the statement of Wilmes' Theorem directly for G.*

Exercise 9.6.5. *Let $G = C_4$ be the four cycle. Sketch the bounded region $\mathcal{B}_G$. Label $\mathcal{B}_G$ to give a cellular resolution of M_G.*

Exercise 9.6.6. *Let $G = K_5$ be the complete graph on 5 vertices. Find the face poset of the bounded region $\mathcal{B}_G$.*

Exercise 9.6.7. *Let $G = C_4$ be the four cycle.*
Compute the kth critical ideal for G for $1 \leq k \leq 4$.

Exercise 9.6.8. *Find the algebraic corank of the cycle graph C_n.*

Exercise 9.6.9. *Let $G = K_4$ be the complete graph on 4 vertices. Determine the following for G:*

1. *The tree ideal M_G.*

2. *The socle elements of M_G.*

3. *The degrees of all socle elements.*

4. *The canonical monomial of M_G.*

Exercise 9.6.10. *Prove that for complete graphs K_n, M_{K_n} satisfies the conditions of the Riemann–Roch Theorem: M_{K_n} is artinian, level and reflection-invariant.*

List of Figures

Bibliography

[AB11]   Arash Asadi and Spencer Backman, *Chip-firing and Riemann–Roch theory for directed graphs*, Electronic Notes in Discrete Mathematics **38** (2011), 63–68.

[AB15]   Omid Amini and Matthew Baker, *Linear series on metrized complexes of algebraic curves*, Mathematische Annalen **362** (2015), no. 1-2, 55–106.

[ABKS14]   Yang An, Matthew Baker, Greg Kuperberg, and Farbod Shokrieh, *Canonical representatives for divisor classes on tropical curves and the matrix-tree theorem*, Forum Math. Sigma **2** (2014), e24, 25. MR 3264262

[AC13]   Omid Amini and Lucia Caporaso, *Riemann–Roch theory for weighted graphs and tropical curves*, Advances in Mathematics **240** (2013), 1–23.

[ACF+16]   Demara Austin, Megan Chambers, Rebecca Funke, Luis David García Puente, and Lauren Keough, *The avalanche polynomial of a graph*, arXiv preprint arXiv:1605.02713 (2016).

[ALS+89]   Richard Anderson, László Lovász, Peter Shor, Joel Spencer, Éva Tardos, and Shmuel Winograd, *Disks, balls, and walls: analysis of a combinatorial game*, Amer. Math. Monthly **96** (1989), no. 6, 481–493. MR 999411

[AM10]   Omid Amini and Madhusudan Manjunath, *Riemann–Roch for sub-lattices of the root lattice A_n*, Electron. J. Combin **17** (2010), no. 1.

[AV14]   Carlos A. Alfaro and Carlos E. Valencia, *Graphs with two trivial critical ideals*, Discrete Applied Mathematics **167** (2014), 33–44.

[Bac17]   Spencer Backman, *Riemann–Roch theory for graph orientations*, Adv. Math. **309** (2017), 655–691. MR 3607288

[Bak96]   Per Bak, *How nature works*, Copernicus, New York, 1996, The science of self-organized criticality. MR 1417042

[Bak08]   Matthew Baker, *Specialization of linear systems from curves to graphs*, Algebra Number Theory **2** (2008), no. 6, 613–653, With an appendix by Brian Conrad. MR 2448666

[BB05]   Anders Björner and Francesco Brenti, *Combinatorics of Coxeter groups*, Graduate Texts in Mathematics, vol. 231, Springer, New York, 2005. MR 2133266

[BBY17]   Spencer Backman, Matthew Baker, and Chi Ho Yuen, *Geometric bijections for regular matroids, zonotopes, and Ehrhart theory*, arXiv preprint arXiv:1701.01051 (2017).

[BCC+17]   Benjamin Braun, Hugo Corrales, Scott Corry, Luis David García Puente, Darren Glass, Nathan Kaplan, Jeremy L. Martin, Gregg Musiker, and Carlos E. Valencia, *Counting arithmetical structures on paths and cycles*, arXiv preprint arXiv:1701.06377 (2017).

[BCFV03]   Franco Bagnoli, Fabio Cecconi, Alessandro Flammini, and Alessandro Vespignani, *Short-period attractors and nonergodic behavior in the deterministic fixed-energy sandpile model*, EPL (Europhysics Letters) **63** (2003), no. 4, 512.

[BCT10]   Brian Benson, Deeparnab Chakrabarty, and Prasad Tetali, *G-parking functions, acyclic orientations and spanning trees*, Discrete Math. **310** (2010), no. 8, 1340–1353. MR 2592488 (2011i:05152)

[BdlHN97]   Roland Bacher, Pierre de la Harpe, and Tatiana Nagnibeda, *The lattice of integral flows and the lattice of integral cuts on a finite graph*, Bull. Soc. Math. France **125** (1997), no. 2, 167–198. MR 1478029 (99c:05111)

[Ber08]   Olivier Bernardi, *Tutte polynomial, subgraphs, orientations and sandpile model: new connections via embeddings*, Electron. J. Combin. **15** (2008), no. 1, Research Paper 109, 53. MR 2438581

[BF06]   Matthew Baker and Xander Faber, *Metrized graphs, Laplacian operators, and electrical networks*, Contemporary Mathematics **415** (2006), no. 15-34, 2.

[BG92]   Javier Bitar and Eric Goles, *Parallel chip firing games on graphs*, Theoretical Computer Science **92** (1992), no. 2, 291–300.

[Big97]   Norman Biggs, *Algebraic potential theory on graphs*, Bulletin of the London Mathematical Society **29** (1997), no. 6, 641–682.

[Big99a]   N. L. Biggs, *Chip-firing and the critical group of a graph*, J. Algebraic Combin. **9** (1999), no. 1, 25–45. MR 1676732 (2000h:05103)

[Big99b]   Norman Biggs, *The Tutte polynomial as a growth function*, Journal of Algebraic Combinatorics **10** (1999), no. 2, 115–133.

[BKR16]   Georgia Benkart, Caroline Klivans, and Victor Reiner, *Chip firing on Dynkin diagrams and McKay quivers*, Mathematische Zeitschrift (2016), 1–34.

[BL92]   Anders Björner and László Lovász, *Chip-firing games on directed graphs*, J. Algebraic Combin. **1** (1992), no. 4, 305–328. MR 1203679

[BL02]   Siegfried Bosch and Dino Lorenzini, *Grothendiecks pairing on component groups of Jacobians*, Inventiones mathematicae **148** (2002), no. 2, 353–396.

[BL16a]   Benjamin Bond and Lionel Levine, *Abelian networks I. Foundations and examples*, SIAM Journal on Discrete Mathematics **30** (2016), no. 2, 856–874.

[BL16b]   ———, *Abelian networks II: halting on all inputs*, Selecta Mathematica **22** (2016), no. 1, 319–340.

[BL16c]   ———, *Abelian networks III: The critical group*, Journal of Algebraic Combinatorics **43** (2016), no. 3, 635–663.

[BLS91]   Anders Björner, László Lovász, and Peter W. Shor, *Chip-firing games on graphs*, European J. Combin. **12** (1991), no. 4, 283–291. MR 1120415 (92g:90193)

[BMM+12]   Andrew Berget, Andrew Manion, Molly Maxwell, Aaron Potechin, and Victor Reiner, *The critical group of a line graph*, Ann. Comb. **16** (2012), no. 3, 449–488. MR 2960015

[BN07]   Matthew Baker and Serguei Norine, *Riemann–Roch and Abel–Jacobi theory on a finite graph*, Adv. Math. **215** (2007), no. 2, 766–788. MR 2355607 (2008m:05167)

[BN09]   _____, *Harmonic morphisms and hyperelliptic graphs*,
              International Mathematics Research Notices **2009** (2009),
              no. 15, 2914–2955.

[Bol76]   Ethan D. Bolker, *Simplicial geometry and transportation
              polytopes*, Transactions of the American Mathematical So-
              ciety **217** (1976), 121–142.

[BR15]   Matthias Beck and Sinai Robins, *Computing the continu-
              ous discretely*, second ed., Undergraduate Texts in Math-
              ematics, Springer, New York, 2015, Integer-point enumer-
              ation in polyhedra, With illustrations by David Austin.
              MR 3410115

[BS13]   Matthew Baker and Farbod Shokrieh, *Chip-firing games,
              potential theory on graphs, and spanning trees*, J. Combin.
              Theory Ser. A **120** (2013), no. 1, 164–182. MR 2971705

[BTW88]   Per Bak, Chao Tang, and Kurt Wiesenfeld, *Self-organized
              criticality*, Phys. Rev. A (3) **38** (1988), no. 1, 364–374.
              MR 949160 (89g:58126)

[BW97]   Norman Biggs and Peter Winkler, *Chip-firing and the
              chromatic polynomial*, Tech. Report LSE-CDAM-97-03,
              London School of Economics, Center for Discrete and Ap-
              plicable Mathematics, 1997.

[BW17]   Matthew Baker and Yao Wang, *The Bernardi process and
              torsor structures on spanning trees*, International Mathe-
              matics Research Notices (2017), rnx037.

[Cap13]   Lucia Caporaso, *Rank of divisors on graphs: an algebro-
              geometric analysis*, A celebration of algebraic geometry,
              Clay Math. Proc., vol. 18, Amer. Math. Soc., Providence,
              RI, 2013, pp. 45–64. MR 3114936

[Car13]   Dustin Cartwright, *Tropical complexes*, arXiv preprint
              arXiv:1308.3813 (2013).

[Cas97]   J. W. S. Cassels, *An introduction to the geometry of num-
              bers*, Classics in Mathematics, Springer-Verlag, Berlin,
              1997, Corrected reprint of the 1971 edition. MR 1434478

[CCG15]   Melody Chan, Thomas Church, and Joshua A. Grochow,
              Rotor-routing and spanning trees on planar graphs, Int.
              Math. Res. Not. IMRN (2015), no. 11, 3225–3244. MR
              3373049

[CDPR12]   Filip Cools, Jan Draisma, Sam Payne, and Elina Robeva, *A tropical proof of the Brill–Noether theorem*, Advances in Mathematics **230** (2012), no. 2, 759–776.

[CDR04]   Robert Cori, Arnaud Dartois, and Dominique Rossin, *Avalanche polynomials on some families of graphs*, Trends in Mathematics (2004), 81–94.

[CE02]   Fan Chung and Robert B. Ellis, *A chip-firing game and dirichlet eigenvalues*, Discrete Mathematics **257** (2002), no. 2-3, 341–355.

[Cha18]   Swee Hong Chan, *Abelian sandpile model and Biggs–Merino polynomial for directed graphs*, Journal of Combinatorial Theory, Series A **154** (2018), 145–171.

[CL87]   Raul Cordovil and Bernt Lindström, *Simplicial matroids*, Combinatorial geometries, Encyclopedia Math. Appl., vol. 29, Cambridge Univ. Press, Cambridge, 1987, pp. 98–113. MR 921070

[CLB03]   Robert Cori and Yvan Le Borgne, *The sand-pile model and Tutte polynomials*, Adv. in Appl. Math. **30** (2003), no. 1-2, 44–52, Formal power series and algebraic combinatorics (Scottsdale, AZ, 2001). MR 1979782

[CLB16]   ———, *On computation of Baker and Norine's rank on complete graphs*, The Electronic Journal of Combinatorics **23** (2016), no. 1, P1–31.

[CLP15]   Julien Clancy, Timothy Leake, and Sam Payne, *A note on Jacobians, Tutte polynomials, and two-variable zeta functions of graphs*, Experimental Mathematics **24** (2015), no. 1, 1–7.

[Cor10]   Scott Corry, *Genus bounds for harmonic group actions on finite graphs*, International Mathematics Research Notices **2011** (2010), no. 19, 4515–4533.

[CP05]   Denis Chebikin and Pavlo Pylyavskyy, *A family of bijections between G-parking functions and spanning trees*, J. Combin. Theory Ser. A **110** (2005), no. 1, 31–41. MR 2128964 (2005m:05010)

[CPS08]   Sergio Caracciolo, Guglielmo Paoletti, and Andrea Sportiello, *Explicit characterization of the identity configuration in an abelian sandpile model*, J. Phys. A **41** (2008), no. 49, 495003, 17. MR 2515893

[CR36]   Alonzo Church and J. Barkley Rosser, *Some properties of conversion*, Transactions of the American Mathematical Society **39** (1936), no. 3, 472–482.

[CR00]   Robert Cori and Dominique Rossin, *On the sandpile group of dual graphs*, European Journal of Combinatorics **21** (2000), no. 4, 447–459.

[CR02]   Hans Christianson and Victor Reiner, *The critical group of a threshold graph*, Linear Algebra Appl. **349** (2002), 233–244. MR 1903735 (2003d:05102)

[Cre04]   Michael Creutz, *Playing with sandpiles*, Physica A: Statistical Mechanics and its Applications **340** (2004), no. 4, 521 – 526, Complexity and Criticality: in memory of Per Bak (1947–2002).

[CRS02]   Robert Cori, Dominique Rossin, and Bruno Salvy, *Polynomial ideals for sandpiles and their Gröbner bases*, Theoret. Comput. Sci. **276** (2002), no. 1-2, 1–15. MR 1896344

[CV10]   Lucia Caporaso and Filippo Viviani, *Torelli theorem for graphs and tropical curves*, Duke Math. J. **153** (2010), no. 1, 129–171.

[CV13]   Hugo Corrales and Carlos E. Valencia, *On the critical ideals of graphs*, Linear Algebra and its Applications **439** (2013), no. 12, 3870–3892.

[CV15]   Hugo Corrales and Carlos E. Valencia, *Critical ideals of trees*, arXiv preprint arXiv:1504.06239 (2015).

[CV18]   ———, *Arithmetical structures on graphs*, Linear Algebra Appl. **536** (2018), 120–151. MR 3713448

[Dav08]   Michael W. Davis, *The geometry and topology of Coxeter groups*, London Mathematical Society Monographs Series, vol. 32, Princeton University Press, Princeton, NJ, 2008. MR 2360474

[DF91]   Persi Diaconis and William Fulton, *A growth model, a game, an algebra, lagrange inversion, and characteristic classes*, Rend. Sem. Mat. Univ. Pol. Torino **49** (1991), no. 1, 95–119.

[Dha90]   Deepak Dhar, *Self-organized critical state of sandpile automaton models*, Phys. Rev. Lett. **64** (1990), no. 14, 1613–1616. MR 1044086 (90m:82053)

[Dha06]   Deepak Dhar, *Theoretical studies of self-organized criti-cality*, Physica A: Statistical Mechanics and its Applica-tions **369** (2006), no. 1, 29–70, Proceedings of the 11th International Summer School on Fundamental Problems in Statistical Physics, September 4–17, 2005, Leuven, Bel-gium.

[DKM09]   Art M. Duval, Caroline J. Klivans, and Jeremy L. Mar-tin, *Simplicial matrix-tree theorems*, Transactions of the American Mathematical Society **361** (2009), no. 11, 6073–6114.

[DKM11]   _____, *The G-shi arrangement, and its relation to G-parking functions*, preprint (2011).

[DKM13]   _____, *Critical groups of simplicial complexes*, Ann. Comb. **17** (2013), no. 1, 53–70. MR 3027573

[DKM15]   _____, *Cuts and flows of cell complexes*, J. Algebraic Comb. **41** (2015), no. 4, 969–999.

[DKM16]   _____, *Simplicial and cellular trees*, Recent trends in combinatorics, IMA Vol. Math. Appl., vol. 159, Springer, [Cham], 2016, pp. 713–752. MR 3526429

[DLKK12]   Jesús A. De Loera, Yvonne Kemper, and Steven Klee, *h-vectors of small matroid complexes*, Electron. J. Combin. **19** (2012), no. 1, Paper 14, 11. MR 2880645

[dMM01]   Anna de Mier and Criel Merino, *Two non-isomorphic graphs with different critical group and the same Tutte polynomial*, preprint (2001).

[Doc17]   Anton Dochtermann, *Skeleta of G-parking function ideals*, arXiv preprint arXiv:1708.04712 (2017).

[Doc18]   _____, *Spherical parking functions, uprooted trees, and yet another way to count n^n*, arXiv preprint arXiv:1806.04289 (2018).

[DP96]   J. Dodziuk and V.K. Patodi, *Riemannian structures and triangulations of manifolds*, Collected Papers of V.K Pa-todi **40** (1996), 232.

[DR02]   Art Duval and Victor Reiner, *Shifted simplicial com-plexes are Laplacian integral*, Transactions of the Ameri-can Mathematical Society **354** (2002), no. 11, 4313–4344.

[DS14]   Anton Dochtermann and Raman Sanyal, *Laplacian ideals, arrangements, and resolutions*, J. Algebraic Combin. **40** (2014), no. 3, 805–822. MR 3265234

[Eck44]   Beno Eckmann, *Harmonische funktionen und randwertaufgaben in einem komplex*, Commentarii Mathematici Helvetici **17** (1944), no. 1, 240–255.

[EMM11]   Joanna A. Ellis-Monaghan and Criel Merino, *Graph polynomials and their applications I: The Tutte polynomial*, Structural Analysis of Complex Networks, Springer, 2011, pp. 219–255.

[Eri91]   Kimmo Eriksson, *No polynomial bound for the chip firing game on directed graphs*, Proceedings of the American Mathematical Society **112** (1991), no. 4, 1203–1205.

[FK17]   Pedro F. Felzenszwalb and Caroline J. Klivans, *Avalanche sorting*, preprint (2017).

[FL16]   Matthew Farrell and Lionel Levine, *Coeulerian graphs*, Proceedings of the American Mathematical Society **144** (2016), no. 7, 2847–2860.

[FLP10]   Anne Fey, Lionel Levine, and Yuval Peres, *Growth rates and explosions in sandpiles*, Journal of Statistical Physics **138** (2010), no. 1-3, 143–159.

[FMP$^+$14] Laura Florescu, Daniela Morar, David Perkinson, Nick Salter, and Tianyuan Xu, *Sandpiles and dominos*, arXiv preprint arXiv:1406.0100 (2014).

[Gab93]   Andrei Gabrielov, *Avalanches, sandpiles and Tutte decomposition*, The Gelfand Mathematical Seminars, 1990–1992, Springer, 1993, pp. 19–26.

[Gab94]   ———, *Asymmetric abelian avalanches and sandpiles*, preprint (1994).

[Gae16]   Christian Gaetz, *Critical groups of group representations*, Linear Algebra and its Applications **508** (2016), 91–99.

[GHMP17a] Pavel Galashin, Sam Hopkins, Thomas McConville, and Alexander Postnikov, *Root system chip-firing I: Interval-firing*, arXiv preprint arXiv:1708.04850 (2017).

[GHMP17b] ———, *Root system chip-firing II: Central-firing*, arXiv preprint arXiv:1708.04849 (2017).

[Gib10]   Peter Giblin, *Graphs, surfaces and homology*, third ed., Cambridge University Press, Cambridge, 2010. MR 2722281

[Gio07]   Emeric Gioan, *Enumerating degree sequences in digraphs and a cycle–cocycle reversing system*, European Journal of Combinatorics **28** (2007), no. 4, 1351–1366.

[GK08]   Andreas Gathmann and Michael Kerber, *A Riemann–Roch theorem in tropical geometry*, Mathematische Zeitschrift **259** (2008), no. 1, 217–230.

[GK15]   Johnny Guzmán and Caroline Klivans, *Chip-firing and energy minimization on M-matrices*, J. Combin. Theory Ser. A **132** (2015), 14–31. MR 3311336

[GK16]   _____, *Chip firing on general invertible matrices*, SIAM J. Discrete Math. **30** (2016), no. 2, 1115–1127. MR 3504984

[GM97]   Eric Goles and Maurice Margenstern, *Universality of the chip-firing game*, Theoretical Computer Science **172** (1997), no. 1-2, 121–134.

[GM02]   O. Gimenez and Criel Merino, *Two non-isomorphic graphs with different critical group and the same Tutte polynomial*, preprint (2002).

[GM14]   Darren B. Glass and Criel Merino, *Critical groups of graphs with dihedral actions*, European Journal of Combinatorics **39** (2014), 95–112.

[GR01]   Chris Godsil and Gordon Royle, *Algebraic graph theory*, Graduate Texts in Mathematics, vol. 207, Springer-Verlag, New York, 2001. MR 1829620

[GZ83]   Curtis Greene and Thomas Zaslavsky, *On the interpretation of Whitney numbers through arrangements of hyperplanes, zonotopes, non-Radon partitions, and orientations of graphs*, Transactions of the American Mathematical Society **280** (1983), no. 1, 97–126.

[Hat02]   Allen Hatcher, *Algebraic topology*, Cambridge University Press, Cambridge, 2002. MR 1867354

[HLM$^+$08] Alexander E. Holroyd, Lionel Levine, Karola Mészáros, Yuval Peres, James Propp, and David B. Wilson, *Chip-firing and rotor-routing on directed graphs*, In and out of equilibrium. 2, Progr. Probab., vol. 60, Birkhäuser, Basel, 2008, pp. 331–364. MR 2477390 (2010f:82066)

[HLW15] Alexander E. Holroyd, Lionel Levine, and Peter Winkler, *Abelian logic gates*, arXiv preprint arXiv:1511.00422 (2015).

[HMP17] Sam Hopkins, Thomas McConville, and James Propp, *Sorting via chip-firing*, Electron. J. Combin. **24** (2017), no. 3, Paper 3.13, 20. MR 3691530

[HMY12] Christian Haase, Gregg Musiker, and Josephine Yu, *Linear systems on tropical curves*, Mathematische Zeitschrift **270** (2012), no. 3-4, 1111–1140.

[Hop14] Sam Hopkins, *Another proof of Wilmes conjecture*, Discrete Mathematics **323** (2014), 43–48.

[HP16] Sam Hopkins and David Perkinson, *Bigraphical arrangements*, Transactions of the American Mathematical Society **368** (2016), no. 1, 709–725.

[Jár14] Antal A. Járai, *Sandpile models*, arXiv preprint arXiv:1401.0354 (2014).

[Jár18] ———, *The sandpile cellular automaton*, Probabilistic Cellular Automata, Springer, 2018, pp. 79–88.

[JLP15] Daniel C. Jerison, Lionel Levine, and John Pike, *Mixing time and eigenvalues of the abelian sandpile Markov chain*, arXiv preprint arXiv:1511.00666 (2015).

[JM13] Rodney James and Rick Miranda, *A Riemann–Roch theorem for edge-weighted graphs*, Proceedings of the American Mathematical Society **141** (2013), no. 11, 3793–3802.

[JNR03] Brian Jacobson, Andrew Niedermaier, and Victor Reiner, *Critical groups for complete multipartite graphs and Cartesian products of complete graphs*, J. Graph Theory **44** (2003), no. 3, 231–250. MR 2012805 (2004g:05075)

[Kal83] Gil Kalai, *Enumeration of Q-acyclic simplicial complexes*, Israel Journal of Mathematics **45** (1983), no. 4, 337–351.

[Kas61]   Pieter W. Kasteleyn, *The statistics of dimers on a lattice: I. The number of dimer arrangements on a quadratic lattice*, Physica **27** (1961), no. 12, 1209–1225.

[KNTG94]   Marcos A. Kiwi, René Ndoundam, Maurice Tchuente, and Eric Goles, *No polynomial bound for the period of the parallel chip firing game on graphs*, Theoretical Computer Science **136** (1994), no. 2, 527–532.

[KS00]   Motoko Kotani and Toshikazu Sunada, *Jacobian tori associated with a finite graph and its abelian covering graphs*, Adv. in Appl. Math. **24** (2000), no. 2, 89–110. MR 1748964 (2002d:14068)

[KT15]   Viktor Kiss and Lilla Tóthmérész, *Chip-firing games on Eulerian digraphs and-hardness of computing the rank of a divisor on a graph*, Discrete Applied Mathematics **193** (2015), 48–56.

[KW66]   Alan G. Konheim and Benjamin Weiss, *An occupancy discipline and applications*, SIAM Journal on Applied Mathematics **14** (1966), no. 6, 1266–1274.

[LBR02]   Yvan Le Borgne and Dominique Rossin, *On the identity of the sandpile group*, Discrete Mathematics **256** (2002), no. 3, 775–790.

[Lev09]   Lionel Levine, *The sandpile group of a tree*, European J. Combin. **30** (2009), no. 4, 1026–1035. MR 2504661 (2010m:05145)

[Lev11a]   ————, *Parallel chip-firing on the complete graph: devils staircase and Poincaré rotation number*, Ergodic Theory and Dynamical Systems **31** (2011), no. 3, 891–910.

[Lev11b]   ————, *Sandpile groups and spanning trees of directed line graphs*, J. Combin. Theory Ser. A **118** (2011), no. 2, 350–364. MR 2739488 (2012b:05143)

[LLP08]   Itamar Landau, Lionel Levine, and Yuval Peres, *Chip-firing and rotor-routing on $\mathbb{Z}^d$ and on trees*, 20th Annual International Conference on Formal Power Series and Algebraic Combinatorics (FPSAC 2008), Discrete Math. Theor. Comput. Sci. Proc., AJ, Assoc. Discrete Math. Theor. Comput. Sci., Nancy, 2008, pp. 587–598. MR 2721487

288 *Bibliography*

[Lor89]   Dino J. Lorenzini, *Arithmetical graphs*, Math. Ann. **285** (1989), no. 3, 481–501. MR 1019714 (91b:14026)

[Lor91]   ———, *A finite group attached to the Laplacian of a graph*, Discrete Math. **91** (1991), no. 3, 277–282. MR 1129991 (93a:05091)

[Lor00]   Dino Lorenzini, *Arithmetical properties of Laplacians of graphs*, Linear and Multilinear Algebra **47** (2000), no. 4, 281–306. MR 1784872 (2001e:05082)

[Lor08]   ———, *Smith normal form and Laplacians*, Journal of Combinatorial Theory, Series B **98** (2008), no. 6, 1271–1300.

[Lor12]   ———, *Two-variable zeta-functions on graphs and Riemann–Roch theorems*, Int. Math. Res. Not. IMRN (2012), no. 22, 5100–5131. MR 2997050

[LP09]   Lionel Levine and Yuval Peres, *Strong spherical asymptotics for rotor-router aggregation and the divisible sandpile*, Potential Analysis **30** (2009), no. 1, 1.

[LPS16]   Lionel Levine, Wesley Pegden, and Charles K. Smart, *Apollonian structure in the Abelian sandpile*, Geom. Funct. Anal. **26** (2016), no. 1, 306–336. MR 3494492

[LR03]   J. C. Lagarias and E. Rains, *On a two-variable zeta function for number fields*, Ann. Inst. Fourier (Grenoble) **53** (2003), no. 1, 1–68. MR 1973068

[LW95]   László Lovász and Peter Winkler, *Mixing of random walks and other diffusions on a graph*, London Mathematical Society Lecture Note Series (1995), 119–154.

[McD18]   Alex McDonough, *Genus from sandpile torsor algorithm*, arXiv preprint arXiv:1804.07807 (2018).

[MD92]   Satya N. Majumdar and Deepak Dhar, *Equivalence between the abelian sandpile model and the $q \to 0$ limit of the potts model*, Physica A: Statistical Mechanics and its Applications **185** (1992), no. 1-4, 129–145.

[Mer92]   Russell Merris, *Unimodular equivalence of graphs*, Linear Algebra Appl. **173** (1992), 181–189. MR 1170510

[Mer01]   Criel Merino, *The chip firing game and matroid complexes*, Discrete models: combinatorics, computation, and geometry (Paris, 2001), Discrete Math. Theor. Comput. Sci. Proc., AA, Maison Inform. Math. Discrèt. (MIMD), Paris, 2001, pp. 245–255. MR 1888777

[ML97]   Criel Merino López, *Chip firing and the Tutte polynomial*, Ann. Comb. **1** (1997), no. 3, 253–259. MR 1630779

[MNRIVF12] Criel Merino, Steven D. Noble, Marcelino Ramírez-Ibañez, and Rafael Villarroel-Flores, *On the structure of the h-vector of a paving matroid*, European Journal of Combinatorics **33** (2012), no. 8, 1787–1799.

[Moh16]   Fatemeh Mohammadi, *Divisors on graphs, orientations, syzygies, and system reliability*, Journal of Algebraic Combinatorics **43** (2016), no. 2, 465–483.

[Moo70]   J. W. Moon, *Counting labelled trees*, From lectures delivered to the Twelfth Biennial Seminar of the Canadian Mathematical Congress (Vancouver, vol. 1969, Canadian Mathematical Congress, Montreal, Que., 1970. MR 0274333 (43 #98)

[MS05]   Ezra Miller and Bernd Sturmfels, *Combinatorial commutative algebra*, Graduate Texts in Mathematics, vol. 227, Springer-Verlag, New York, 2005. MR 2110098

[MS13]   Madhusudan Manjunath and Bernd Sturmfels, *Monomials, binomials and Riemann–Roch*, J. Algebraic Combin. **37** (2013), no. 4, 737–756. MR 3047017

[MS14]   Fatemeh Mohammadi and Farbod Shokrieh, *Divisors on graphs, connected flags, and syzygies*, Int. Math. Res. Not. IMRN (2014), no. 24, 6839–6905. MR 3291642

[MS16]   ———, *Divisors on graphs, binomial and monomial ideals, and cellular resolutions*, Mathematische Zeitschrift **283** (2016), no. 1, 59–102.

[MSW15]   Madhusudan Manjunath, Frank-Olaf Schreyer, and John Wilmes, *Minimal free resolutions of the G-parking function ideal and the toppling ideal*, Transactions of the American Mathematical Society **367** (2015), no. 4, 2853–2874.

[Mus09]   Gregg Musiker, *The critical groups of a family of graphs and elliptic curves over finite fields*, J. Algebraic Combin. **30** (2009), no. 2, 255–276. MR 2525061 (2010g:14027)

[MZ08]   Grigory Mikhalkin and Ilia Zharkov, *Tropical curves, their Jacobians and theta functions*, Curves and abelian varieties, Contemp. Math., vol. 465, Amer. Math. Soc., Providence, RI, 2008, pp. 203–230. MR 2457739

[New42]   M. H. A. Newman, *On theories with a combinatorial definition of "equivalence"*, Ann. of Math. (2) **43** (1942), 223–243. MR 0007372

[NW11]   Serguei Norine and Peter Whalen, *Jacobians of nearly complete and threshold graphs*, European Journal of Combinatorics **32** (2011), no. 8, 1368–1376.

[Oh13]   Suho Oh, *Generalized permutohedra, h-vectors of cotransversal matroids and pure O-sequences*, The Electronic Journal of Combinatorics **20** (2013), no. 3, P14.

[Ost03]   S. Ostojic, *Patterns formed by addition of grains to only one site of an abelian sandpile*, Physica A Statistical Mechanics and its Applications **318** (2003), 187–199.

[Oxl11]   James Oxley, *Matroid theory*, second ed., Oxford Graduate Texts in Mathematics, vol. 21, Oxford University Press, Oxford, 2011. MR 2849819

[Pao14]   Guglielmo Paoletti, *Deterministic abelian sandpile models and patterns*, Springer Theses, Springer, Cham, 2014, Thesis, University of Pisa, Pisa, 2012. MR 3100415

[Pel96]   Ruud Pellikaan, *On special divisors and the two variable zeta function of algebraic curves over finite fields*, Arithmetic, geometry and coding theory (Luminy, 1993), de Gruyter, Berlin, 1996, pp. 175–184. MR 1394933

[Ple77]   R.J. Plemmons, *M-matrix characterizations. I. Nonsingular M-matrices*, Linear Algebra and Its Applications **18** (1977), no. 2, 175–188.

[PP16]   Kévin Perrot and Trung Van Pham, *Chip-firing game and a partial Tutte polynomial for Eulerian digraphs*, The Electronic Journal of Combinatorics **23** (2016), no. 1, P1–57.

[PPW13]   David Perkinson, Jacob Perlman, and John Wilmes, *Primer for the algebraic geometry of sandpiles*, Tropical and non-Archimedean geometry, Contemp. Math., vol. 605, Amer. Math. Soc., Providence, RI, 2013, pp. 211–256. MR 3204273

[Pri94]   Erich Prisner, *Parallel chip firing on digraphs*, Complex Systems **8** (1994), no. 5, 367–383. MR 1335320

[PS04]   Alexander Postnikov and Boris Shapiro, *Trees, parking functions, syzygies, and deformations of monomial ideals*, Trans. Amer. Math. Soc. **356** (2004), no. 8, 3109–3142 (electronic). MR 2052943

[PS13]   Wesley Pegden and Charles K. Smart, *Convergence of the Abelian sandpile*, Duke Math. J. **162** (2013), no. 4, 627–642. MR 3039676

[PSX11]   David Perkinson, Nick Salter, and Tianyuan Xu, *A note on the critical group of a line graph*, Electron. J. Combin. **18** (2011), no. 1, Paper 124, 6. MR 2811093 (2012d:05180)

[Pyk59]   Ronald Pyke, *The supremum and infimum of the Poisson process*, The Annals of Mathematical Statistics **30** (1959), no. 2, 568–576.

[Sch10]   Jay Schweig, *On the h-vector of a lattice path matroid*, Electron. J. Combin. **17** (2010), no. 1, Note 3, 6. MR 2578897

[Spe86]   J. Spencer, *Balancing vectors in the max norm*, Combinatorica **6** (1986), no. 1, 55–65. MR 856644

[Spe93]   Eugene R. Speer, *Asymmetric abelian sandpile models*, Journal of Statistical Physics **71** (1993), no. 1-2, 61–74.

[Sta96a]   Richard P. Stanley, *Combinatorics and commutative algebra*, second ed., Progress in Mathematics, vol. 41, Birkhäuser Boston, Inc., Boston, MA, 1996. MR 1453579

[Sta96b]   _____, *Hyperplane arrangements, interval orders, and trees*, Proceedings of the National Academy of Sciences **93** (1996), no. 6, 2620–2625.

[Sta99]   _____, *Enumerative combinatorics. Vol. 2*, Cambridge Studies in Advanced Mathematics, vol. 62, Cambridge University Press, Cambridge, 1999, With a foreword by Gian-Carlo Rota and appendix 1 by Sergey Fomin. MR 1676282

[Sta07]   _____, *An introduction to hyperplane arrangements*, Geometric combinatorics, IAS/Park City Math. Ser., vol. 13, Amer. Math. Soc., Providence, RI, 2007, pp. 389–496. MR 2383131

[SW10]   Yi Su and David G. Wagner, *The lattice of integer flows of a regular matroid*, Journal of Combinatorial Theory, Series B **100** (2010), no. 6, 691–703.

[Tar88]   Gábor Tardos, *Polynomial bound for a chip firing game on graphs*, SIAM J. Discrete Math. **1** (1988), no. 3, 397–398. MR 955655

[TF61]   Harold N.V. Temperley and Michael E. Fisher, *Dimer problem in statistical mechanics-an exact result*, Philosophical Magazine **6** (1961), no. 68, 1061–1063.

[Tou07]   Evelin Toumpakari, *On the sandpile group of regular trees*, European J. Combin. **28** (2007), no. 3, 822–842. MR 2300763

[VDGS00]   Gerard Van Der Geer and René Schoof, *Effectivity of arakelov divisors and the theta divisor of a number field*, Selecta Mathematica, New Series **6** (2000), no. 4, 377–398.

[vL90]   Jan van Leeuwen (ed.), *Handbook of theoretical computer science. Vol. B*, Elsevier Science Publishers, B.V., Amsterdam; MIT Press, Cambridge, MA, 1990, Formal models and semantics. MR 1127185

[Wag00]   David G. Wagner, *The critical group of a directed graph*, arXiv preprint math/0010241 (2000).

[Woo17]   Melanie Wood, *The distribution of sandpile groups of random graphs*, Journal of the American Mathematical Society **30** (2017), no. 4, 915–958.

[Yue18]   Chi Ho Yuen, *Geometric bijections of graphs and regular matroids*, Ph.D. thesis, Georgia Institute of Technology, 2018.

Index